R. INGLIS.

Ocean Engineering Wave Mechanics

Ocean Engineering: A Wiley Series

EDITOR: MICHAEL E. McCORMICK, Ph.D.,
U.S. Naval Academy
ASSOCIATE EDITOR:
RAMESWAR BHATTACHARYYA, Dr. Ing.
U.S. Naval Academy

Michael E. McCormick   Ocean Engineering Wave Mechanics

OCEAN ENGINEERING WAVE MECHANICS

MICHAEL E. McCORMICK
Research Professor
Department of Naval Systems Engineering
United States Naval Academy

A WILEY-INTERSCIENCE PUBLICATION

JOHN WILEY & SONS
New York • Chichester • Brisbane • Toronto

Copyright © 1973, by John Wiley & Sons, Inc.

All rights reserved. Published simultaneously in Canada.

Reproduction or translation of any part of this work beyond that permitted by Sections 107 or 108 of the 1976 United States Copyright Act without the permission of the copyright owner is unlawful. Requests for permission or further information should be addressed to the Permissions Department, John Wiley & Sons, Inc.

Library of Congress Cataloging in Publication Data:

McCormick, Michael E   1936–
 Ocean engineering wave mechanics.

(Ocean engineering: A Wiley Series)
"A Wiley-Interscience publication."
Includes bibliographies.
1. Ocean engineering. 2. Hydraulic structures.
3. Hydrodynamics. I. Title.

TC1645.M32   620′.416′2   72-12756
ISBN 0-471-58177-1

Printed in the United States of America

10 9 8 7 6 5 4

To Mary Ann and the children for tolerating me during the preparation of the manuscript, and to Dr. David A. Jewell and Professor Bernard Morrill for their help and encouragement during my formative years as a hydrodynamicist and teacher.

SERIES PREFACE

Ocean engineering is both old and new. It is old in that man has concerned himself with specific problems in the ocean for thousands of years. Ship building, prevention of beach erosion, and construction of offshore structures are just a few of the specialties that have been developed by engineers over the ages. Until recently, however, these efforts tended to be restricted to specific areas. Within the past decade an attempt has been made to coordinate the activities of all technologists in ocean work, calling the entire field "ocean engineering." Here we have its newness.

Ocean Engineering: A Wiley Series has been created to introduce engineers and scientists to the various areas of ocean engineering. Books in this series are so written as to enable engineers and scientists easily to learn the fundamental principles and techniques of a specialty other than their own. The books can also serve as text books in advanced undergraduate and introductory graduate courses. The topics to be covered in this series include ocean engineering wave mechanics, marine corrosion, coastal engineering, dynamics of marine vehicles, offshore structures, and geotechnical or seafloor engineering. We think that this series fills a great need in the literature of ocean technology.

MICHAEL E. MCCORMICK, EDITOR
RAMESWAR BHATTACHARYYA, ASSOCIATE EDITOR

November 1972

PREFACE

The purpose of this book is to acquaint the reader with the analytical and experimental techniques used in solving mechanical problems involving surface waves. Specifically, the surface wave and its effects on fixed and floating structures are discussed.

The book evolved from the class notes used in a one-semester senior-postgraduate level course entitled "Ocean Engineering Mechanics," taught at the U.S. Naval Academy. The students taking this course have an educational background that includes courses in solid and fluid mechanics, ordinary and partial differential equations, and vector analysis. The book is designed to serve as both a classroom and a self-educational text. By reading the material contained herein, the engineer, physicist, or mathematician entering the ocean engineering field can easily learn to apply his talents to problems involving ocean wave mechanics.

The course from which this book results is preceded at the U.S. Naval Academy by a survey-type course in ocean engineering, as well as a course covering ocean engineering structures and materials. It is followed by a course entilted "Advanced Marine Vehicles," which is taken as an elective by students majoring in the fields of naval architecture and ocean engineering. The emphasis is, therefore, on technique rather than specific problem solving. Where appropriate, however, specific situations are discussed.

I wish to express my sincere appreciation to Dr. R. Bhattacharyya for his advice and encouragement. In addition, special thanks are due to Dr. Bruce Johnson, Dr. N. T. Monney, and Lieutenant Commander J. W. Eckert, as well as to my former students for their fine editorial comments and their interest.

MICHAEL E. MCCORMICK

Annapolis, Maryland
September 1972

CONTENTS

Notation   xv

1. Review of Hydromechanics   1

1.1 *Hydrostatics*, 1
1.2 *Equation of Continuity*, 4
1.3 *Rotational and Irrotational Flows*, 5
1.4 *The Dynamical Equations of Motion*, 9
1.5 *Viscous Flows*, 11
1.6 *Illustrative Examples*, 14
      a. Collapse Depth, 14
      b. Hydrofoil Lift, 16
      c. Drag on Cylinders and Reynolds Scaling, 19
1.7 *Summary*, 21
1.8 *References*, 21
1.9 *Problems*, 21

2. Surface Waves   24

2.1 *Linear Wave Theory*, 25
2.2 *The Wave Group*, 36
2.3 *Wave Energy*, 37
2.4 *Nonlinear Waves*, 40
      A. Stokes's Theory, 41
      B. Gerstner's "Trochoidal" Theory, 48

2.5 *Experimental Wave Study, 50*

    a. Wave Measurement, 50
    b. Wave Experiment, 52

2.6 *Illustrative Examples, 54*

    a. Circular Surface Waves, 54
    b. Interfacial Waves, 57
    c. A Computer Solution of Wavelength by Successive Approximations, 59

2.7 *Summary, 61*
2.8 *References, 61*
2.9 *Problems, 62*

3. Fixed Structures in Waves   64

3.1 *Hydrostatic Pressure Beneath a Surface Wave, 64*
3.2 *Waves at a Vertical Flat Barrier, 67*
3.3 *Waves on a Sloping Flat Barrier, 73*
3.4 *Consequences of Viscosity, 77*
3.5 *Wave-Induced Forces on a Pile, 78*
3.6 *Wave-Induced Vibrations of Fixed Structures, 85*
3.7 *Wave-Making Drag, 89*
3.8 *Experimental Studies, 95*

    a. Wave-Induced Forces and Motions, 95
    b. Wave-Making Resistance, 98

3.9 *Illustrative Examples, 101*

    a. Wave-Induced Forces and Moments, 101
    b. Shoaling Waves and Wave Refraction: Computer Solution, 103
    c. Wave-Making Resistance, 107

3.10 *References, 109*
3.11 *Problems, 110*

4. Floating Structures in Waves — 112

4.1 Coupled Heaving and Pitching, 112

    A. *Motion-Induced Force and Moment, 116*
    B. *Hydrostatic Restoring Force and Moment, 120*
    C. *Damping, 121*
    D. *Wave-Induced Force and Moment, 124*
    E. *Equations of Motion of Heave and Pitch, 129*

4.2 *Moored and Towed Bodies, 137*
4.3 *Experimental Study of Body Motions, 147*
4.4 *Illustrative Examples, 152*

    a. Coupled Heaving and Pitching of a Fully Submerged Body, 152
    b. The Semisubmerged Hull: Effective Spring Constant of a Mooring Cable, 158
    c. Computer Analysis of Uncoupled Buoy Motions, 163

4.5 *References, 165*
4.6 *Problems, 166*

Appendices — 169

A. *FORTRAN IV Language, 169*
B. *Shoaling Wave Tables, 171*
C. *Answers to Selected Problems 173*

Index — 177

NOTATION

COMMON TO ALL CHAPTERS

a	wave amplitude, ft
A	area vector $(A_x\mathbf{i} + A_y\mathbf{j} + A_z\mathbf{k})$, ft²
c	celerity or phase velocity, fps
c_g	group velocity, fps
C_j	force coefficient
D	diameter, ft
D	drag, lb
E	energy per wavelength, lb-ft/ft
f	frequency, Hz
F	force, lb
g	gravitational acceleration, ft/sec²
h	water depth, ft
H	wave height, ft
i, **j**, **k**	unit vectors in the x, y, z directions
k	wave number $(2\pi/\lambda)$, 1/ft
L	length, ft
m	mass, slugs
m_w	added-mass, slugs/ft
M	moment, lb/ft
n	normal unit vector to the free surface
N	normal unit vector to the sea floor
p	pressure, psf
r	position vector $(\xi\mathbf{i} + \zeta\mathbf{k})$, ft

xvi   NOTATION

R	radius, ft
R_D	Reynolds number
S	Strouhal number
S.W.L.	Still-water level
t	time, sec
T	wave period $(1/f)$, sec
u, v, w	velocity components in the x, y, z directions
v	volume, ft^3
V	velocity vector $(\lambda\mathbf{i} + v\mathbf{j} + w\mathbf{k})$, fps
W	weight, lb
x, y, z	Cartesian coordinates
x_0, z_0	origin of ξ, ζ
γ	weight density, lb/ft^3
$\eta(t)$	free-surface displacement, ft
λ	wavelength, ft
μ	dynamic viscosity, lb-sec/ft^2
ν	kinematic viscosity, ft^2/sec
ξ, ζ	coordinates with origins at x_0, z_0
ρ	mass density, slugs/ft^3
φ	velocity potential, ft^2/sec
ψ	stream function, ft^2/sec
ω	circular frequency $(2f)$, 1/sec

Subscripts

d	divergent
i	inertial
n	normal
0	deep-water value
R	wave
t	transverse
T	total

CHAPTER 1 NOTATION

b	hydrofoil span, ft
c	hydrofoil chord, ft

h	wall thickness, ft
L	lift, lb
S	integration path
T	cable tension, lb
α	elevation angle
Γ	circulation, ft²/sec
σ	normal stress, psf
τ	shear stress, psf

CHAPTER 2 NOTATION

J_m	Bessel function of the first kind of order m
r, β, z	polar coordinates
U_{con}	convection velocity, fps
θ	slope of the free surface
ξ	dummy variable in Eq. 2.44
Ω	vorticity, 1/sec

CHAPTER 3 NOTATION

$a_{1,2,3}$	modal constants
$A(\theta)$	amplitude function in Eq. 3.39, ft
C	wave resistance parameter in Eq. 3.41, ft
E	Young's modulus, psf
I	area moment of inertia, ft⁴
P	dimensionless pressure in Eq. 3.5b
Z	dimensionless depth (z/λ)
α	beach angle
β	refraction angle
θ	horizontal angle in Eq. 3.39; a dummy variable in Eq. 3.13

CHAPTER 4 NOTATION

a, b, c, d, e, h	hydrodynamic coefficients for heave; see Eq. 4.38
a_d	amplitude of the damping wave, ft

xviii   NOTATION

A, B, C, D, E, H	hydrodynamic coefficients for pitch; see Eq. 4.40
$\bar{A}$	damping amplitude ratio (a_d/A_b)
A_b	amplitude of the body motion, ft
B'	beam at the water line, ft
$B'(\xi)$	sectional beam at the water line, ft
d	depth, ft
E_c	Young's modulus, psi
$f_{ct,cn}$	see Eqs. 4.65 and 4.66
$F_{A,B,0}$	force coefficients in Eq. 4.38, lb
G	center of gravity
I_y	moment of inertia about y axis, lb-ft-sec^2
k_c	spring constant of elastic cable, lb/ft
L_{11}, L_{12}, etc.	differential operaters in Eq. 4.43
L	length, ft
$m_w(\xi)$	added-mass per unit length, lb-sec^2/ft^2
$M_{A,B,0}$	moment coefficients in Eq. 4.40, lb/ft
N	damping force per unit vertical velocity, lb-sec/ft
r	radius of curvature, ft
R	cable radius, ft
s	curvilinear coordinate, ft
T	cable tension, lb
T'_d	draft, ft
Z, Z_θ	magnification factors in Eqs. 4.52 and 4.56
α	angular body coordinate in Figures 4.7 and 4c.
β	angular body coordinate in Figure 4c
γ	phase angle; see Eq. 4.38
δ	phase angle, see Eq. 4.40
Δ	damping ratio; see Eqs. 4.50 and 4.54
ε	strain
ζ	vertical displacement, ft
θ	rotational displacement, rad
μ	Poisson's ratio
σ	phase angle; see Eq. 4.49
φ	cable angle, rad
ψ	buoy angle, rad

Subscripts

a	at anchor
b	of or on body
c	cable
d	damping
e	encounter
L	lower
n	natural frequency or normal direction
0	at float or constant value
r	radial
s	axial
U	upper
θ	pitching

Ocean Engineering Wave Mechanics

Chapter ONE

REVIEW OF HYDROMECHANICS

The study of free-surface waves and their effects on fixed and floating structures requires a prior knowledge of fluid mechanics. Therefore the first chapter of this book is devoted to a review of the fundamentals of hydromechanics. The material presented in this chapter is essentially that which is used in later derivations and analyses. Since the purpose of the chapter is to review the material, the various topics are discussed in a rather cursory manner. The reader is referred to the books by Shames (1962) and by Schlichting (1960) for more thorough coverage.

1.1. HYDROSTATICS

One of the major problems encountered in placing man in the sea is that of dealing with the severe stresses on both man and his habitat and vehicle resulting from the hydrostatic pressures encountered at moderate and extreme ocean depths. Hence a review of the fundamentals of hydrostatics is presented in this section.

2 REVIEW OF HYDROMECHANICS

Consider an elemental volume of water, ΔV, at a depth, h, below the *free surface (air-sea interface)*, as shown in Figure 1.1. Let the volume be bounded by vertical lines of average length Δz and by the area elements $\Delta \mathbf{A}_1$ and $\Delta \mathbf{A}_2$. If the element is assumed to be in a state of static equilibrium, the force equation in the vertical or z direction is

$$\left[-\left(p + \frac{\partial p}{\partial z}\Delta z\right)\Delta \mathbf{A}_1 - p\,\Delta \mathbf{A}_2 - \gamma\,\Delta V\,\mathbf{k} \right] \cdot \mathbf{k} = 0$$

where $\mathbf{i}$, $\mathbf{j}$, and $\mathbf{k}$ are the unit vectors in the x, y, and z directions, respectively, and γ is the specific weight of sea water. *Note:* The included angle between the area vector $\Delta \mathbf{A}_2$ and $\mathbf{k}$ is greater than $\pi/2$; hence the scalar product $\Delta \mathbf{A}_2 \cdot \mathbf{k}$ is negative. The projections of the elemental areas onto a bisecting horizontal plane are equal to

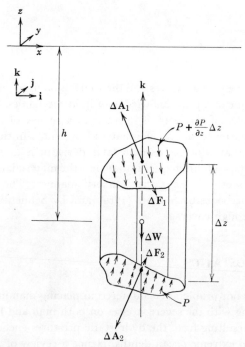

Figure 1.1. A fluid element in static equilibrium.

1.1. HYDROSTATICS

the cross-sectional area of the fluid column, that is, $|\Delta \mathbf{A}_1 \cdot \mathbf{k}| = |\Delta \mathbf{A}_2 \cdot \mathbf{k}| = \Delta A_z$; therefore, the equilibrium equation reduces to the following:

$$\frac{\partial p}{\partial z} = -\gamma \qquad (1.1)$$

Equation 1.1 is called the *hydrostatic equation*. Applying the same analysis to a fluid element bounded by horizontal lines, one finds that the pressure gradients in the x and y directions are zero, that is,

$$\frac{\partial p}{\partial x} = \frac{\partial p}{\partial y} = 0 \qquad (1.2)$$

Thus the *hydrostatic theorem* is proved: The pressure acting on a plane normal to the direction of the force field (in this case the gravitational direction) is constant in a continuous fluid.

The hydrostatic equation is, from Eq. 1.2, an ordinary differential equation and can be integrated directly. Hence Eq. 1.1 can be written as

$$\frac{dp}{dz} = -\gamma \qquad (1.3)$$

from which the hydrostatic pressure is found to be

$$p = -\int_0^{-h} \gamma \, dz = \begin{cases} \gamma h, & \gamma \text{ invariant} \\ f(h), & \gamma \text{ variable} \end{cases} \qquad (1.4)$$

One learns in an elementary course in fluid mechanics that water is an incompressible fluid. At the extreme depths of the ocean, however, sea water is compressed by the immense pressures encountered. If the specific weight of the sea water is assumed to vary linearly from 64 lb/ft³ at the free surface to 66.6 lb/ft³ at a depth of 30,000 ft, as King (1969) assumes, then the hydrostatic pressure variation in the ocean is given by

$$p = -\int_0^{-h} (64 - 0.866 \times 10^{-4} z) \, dz$$

$$= 64h + (0.433 \times 10^{-4}) h^2 \qquad (1.5)$$

4 REVIEW OF HYDROMECHANICS

By using Eq. 1.5, the pressure at 30,000 ft is found to be 1.959×10^6 lb/ft^2, whereas the pressure calculated by neglecting the compressibility is 1.920×10^6 lb/ft^2. Thus the inclusion of the compressibility effect results in an approximate 2% increase in the calculated pressure at this depth. Since most structural designers use safety factors of 100% or more, consideration of the compressibility of sea water in habitat or vehicle design is somewhat superfluous.

The compressibility of sea water must also be accounted for when considering buoyancy in the deep ocean. For instance, a neutrally buoyant body displacing 1 ft^3 of sea water near the free surface will have a buoyant force of 2.6 lb acting on it at a depth of 30,000 ft, since 1 ft^3 of water at that depth weighs 66.6 lb.

1.2. EQUATION OF CONTINUITY

The fact that a fluid can neither be created nor destroyed is expressed mathematically by the equation of continuity. To derive this equation consider a fluid flowing through an elemental volume ΔV fixed in space as shown in Figure 1.2. At the point (x, y, z) the mass

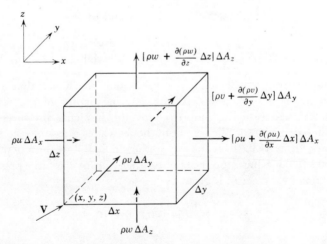

Figure 1.2. Conservation of mass.

1.3. ROTATIONAL AND IRROTATIONAL FLOWS

density of the fluid is ρ, and the velocity is $\mathbf{V} = u\mathbf{i} + v\mathbf{j} + w\mathbf{k}$. The law of the conservation of mass dictates that the internal rate of mass accumulation must be equal to the net rate of mass influx across the surfaces of the element. Mathematically this law is expressed by

$$\frac{\partial(\rho \, \Delta V)}{\partial t} = -\left\{ \frac{\partial(\rho u)}{\partial x} \Delta x \, \Delta A_x + \frac{\partial(\rho v)}{\partial y} \Delta y \, \Delta A_y + \frac{\partial(\rho w)}{\partial z} \Delta z \, \Delta A_z \right\}$$

Since $\Delta V = \Delta x \, \Delta A_x = \Delta y \, \Delta A_y = \Delta z \, \Delta A_z$, the conservation of mass equation reduces to the following:

$$\frac{\partial \rho}{\partial t} = -\left[\frac{\partial(\rho u)}{\partial x} + \frac{\partial(\rho v)}{\partial y} + \frac{\partial(\rho w)}{\partial z} \right] \quad (1.6)$$

$$= -\Delta \cdot (\rho \mathbf{V})$$

Equation 1.6 is called the *equation of continuity*. For a steady flow, both compressible and incompressible, Eq. 1.6 reduces to

$$\nabla \cdot (\rho \mathbf{V}) = 0 \quad (1.7)$$

If the flow is incompressible, both steady and unsteady, the continuity equation is

$$\nabla \cdot \mathbf{V} = 0 \quad (1.8)$$

Most of the problems in ocean engineering fluid dynamics can be solved by assuming the flow to be incompressible. There are special situations, however, in which density variations must be considered. For example, the cooling water coming from a nuclear reactor will be considerably warmer than the river, bay, or ocean water into which it empties. Hence a current will result because of the buoyancy of the lighter warm water.

1.3. ROTATIONAL AND IRROTATIONAL FLOWS

A rather important concept in hydrodynamics is *circulation*. Among other applications, this quantity is used to determine the theoretical lift of a hydrofoil. Circulation is defined as the line integral of the tangent velocity of a fluid about a closed path S, referring to

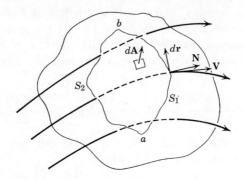

Figure 1.3. Circulation in three dimensions.

Figure 1.3. Using Stokes's integral theorem, as in Newell (1955), one can write the circulation as

$$\Gamma = \oint \mathbf{V} \cdot d\mathbf{r} = \oiint \nabla \times \mathbf{V} \cdot d\mathbf{A} \qquad (1.9)$$

where $d\mathbf{A}$ is the differential of the area of the surface on which the curve S is situated. If the line integration is independent of the path between any two points a and b on S, then the circulation is zero since

$$\int_a^b \mathbf{V} \cdot d\mathbf{r} \bigg|_{S_1} = -\int_b^a \mathbf{V} \cdot d\mathbf{r} \bigg|_{S_2}$$

In the remaining portion of this section, as well as in the section to follow, advantage is taken of the fact that the scalar product of the gradient of a function and a differential of a direction vector results in the differential of the function in question; that is, if E is a continuous function and $d\mathbf{r} = dx\,\mathbf{i} + dy\,\mathbf{j} + dz\,\mathbf{k}$, then

$$\nabla E \cdot d\mathbf{r} = dE$$

By using the previous equation, the line integral in Eq. 1.9 can be expressed in terms of the scalar function ϕ as

$$\oint \mathbf{V} \cdot d\mathbf{r} = \oint d\phi = \oint \nabla \phi \cdot d\mathbf{r}$$

Comparing the first and the last integrands, one obtains the relation

$$\mathbf{V} = \nabla \phi \qquad (1.10a)$$

1.3. ROTATIONAL AND IRROTATIONAL FLOWS

or

$$u = \frac{\partial \phi}{\partial x}, \text{ etc.} \tag{1.10b}$$

where the scalar function ϕ is called the *velocity potential*. The combination of Eq. 1.10a with the integrand of the surface integral in Eq. 1.9 causes the integral to vanish, since the curl of the gradient of a scalar is identically zero for any continuous scalar function ϕ. From this fact one can conclude that Eq. 1.10 is valid if and only if the circulation is zero. When this is true the flow is said to be *irrotational*, that is, there is no net rotation of the fluid particles about their centers, only distortion of the particles.

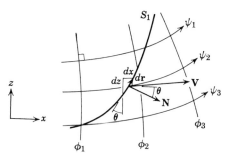

Figure 1.4. Two-dimensional flow.

Consider the flow in Figure 1.4 to be in the plane of the page. The flux of fluid across the line S_1 is

$$\int_a^b \mathbf{V} \cdot \mathbf{N} |d\mathbf{r}| = \int_a^b d\psi = \int_a^b \nabla \psi \cdot d\mathbf{r} \tag{1.11}$$

where $\mathbf{N}$ is the normal unit vector on S_1, and $d\mathbf{r}$ is the elemental vector tangent to the curve. Using the Cartesian coordinate system shown, one can express the vectors in Eq. 1.11 by the following equations:

$$d\mathbf{r} = dx\,\mathbf{i} + dz\,\mathbf{k}$$
$$\mathbf{V} = u\mathbf{i} + w\mathbf{k}$$

and

$$\mathbf{N} = \cos(\theta)\mathbf{i} + \sin(\theta)(-\mathbf{k})$$
$$= \frac{dz}{|d\mathbf{r}|}\mathbf{i} - \frac{dx}{|d\mathbf{r}|}\mathbf{k}$$

8 REVIEW OF HYDROMECHANICS

Substituting these relations into Eq. 1.11 yields the following integral relation:

$$\int_a^b \mathbf{V} \cdot \mathbf{N} |d\mathbf{r}| = \int_a^b u\, dz - \int_a^b w\, dx = \int_a^b \frac{\partial \psi}{\partial x} dx + \int_a^b \frac{\partial \psi}{\partial z} dz$$

or, comparing the terms in the x and z integrals, one obtains

$$u = \frac{\partial \psi}{\partial z}, \qquad w = -\frac{\partial \psi}{\partial x} \qquad (1.12)$$

The scalar function ψ, called the *stream function*, can be used to represent the velocity in both rotational and irrotational two-dimensional flows.

If the flow is irrotational, then by comparing Eqs. 1.10 and 1.12 one obtains

$$u = \frac{\partial \phi}{\partial x} = \frac{\partial \psi}{\partial z}$$

and $\qquad\qquad\qquad\qquad\qquad\qquad\qquad\qquad\qquad\qquad\qquad$ (1.13)

$$w = \frac{\partial \phi}{\partial z} = -\frac{\partial \psi}{\partial x}$$

The equations relating the velocity potential and the stream function are called the *Cauchy-Riemann equations*.

Now consider the geometric curves defined by assigning constant values to both ϕ and ψ. The differentials of these functions along their respective curves are

$$d\phi = \frac{\partial \phi}{\partial x} dx + \frac{\partial \phi}{\partial z} dz = u\, dx + w\, dz = 0$$

or

$$\frac{u}{w} = -\frac{dz}{dx}\bigg|_{\phi \text{ constant}}$$

and

$$d\psi = \frac{\partial \psi}{\partial x} dx = \frac{\partial \psi}{\partial z} dz = -w\, dx + u\, dz = 0$$

or

$$\frac{u}{w} = \frac{dx}{dz}\bigg|_{\psi \text{ constant}}$$

An examination of these results reveals that lines of constants ϕ and ψ form an orthogonal set, as shown in Figure 1.4.

1.4. THE DYNAMICAL EQUATIONS OF MOTION

Newton's second law of motion states that the sum of the external forces acting on a mass must be equal to the time rate of change of linear momentum of the mass. Mathematically, Newton's second law can be stated as

$$\Sigma \mathbf{F} = \frac{d(m\mathbf{V})}{dt} = m\frac{d\mathbf{V}}{dt} + \mathbf{V}\frac{dm}{dt} = m\left(\frac{\partial \mathbf{V}}{\partial t} + \mathbf{V} \cdot \nabla \mathbf{V}\right) \quad (1.14)$$

where $dm/dt = 0$ is a statement of the conservation of mass.

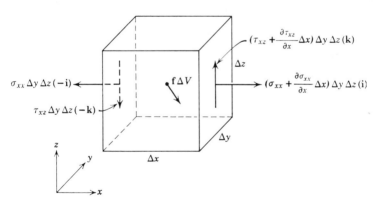

Figure 1.5. External forces acting on a fluid element.

Let us apply this equation to the elemental mass of fluid shown in Figure 1.5, subject to shear stresses, τ_{ij}, normal stresses, σ_{ii}, and a body force, $\mathbf{f}\,\Delta V$. The first subscript in the stress notation refers to the surface upon which the stress acts, and the second subscript designates the direction of the resulting force. If the elemental volume is assumed to be constant, Eq. 1.14 applied to the fluid is

$$\rho \Delta V \left(\frac{\partial \mathbf{V}}{\partial t} + \mathbf{V} \cdot \nabla \mathbf{V} \right) = \mathbf{f} \Delta V +$$

$$\left[\left(\frac{\partial \sigma_{xx}}{\partial x} + \frac{\partial \tau_{yx}}{\partial y} + \frac{\partial \tau_{zx}}{\partial z} \right) \mathbf{i} + \left(\frac{\partial \tau_{xy}}{\partial x} + \frac{\partial \sigma_{yy}}{\partial y} + \frac{\partial \tau_{zy}}{\partial z} \right) \mathbf{j} + \left(\frac{\partial \tau_{xz}}{\partial x} + \frac{\partial \tau_{yz}}{\partial y} + \frac{\partial \sigma_{zz}}{\partial z} \right) \mathbf{k} \right] \quad (1.15)$$

The tensile stress in a fluid is composed of two components: the hydrostatic pressure, p, which is a negative stress, and a component, $\bar{\sigma}_{ii}$, proportional to the time rate of change of strain. Therefore the normal stresses in Eq. 1.15 are

$$\sigma_{ii} = -p + \bar{\sigma}_{ii} \quad (1.16)$$

where [Schlichting (1960)]

$$\bar{\sigma}_{ii} = 2\mu \frac{\partial u_i}{\partial x_i} \quad (1.17)$$

the single subscript referring to the velocity and coordinate direction. The coefficient, μ, of the derivative in Eq. 1.17 is called the *coefficient of viscosity*.

The shear stresses in Eq. 1.15 [Schlichting (1960)] can also be expressed in terms of rates of change of strain as

$$\tau_{ij} = \tau_{ji} = \mu \left(\frac{\partial u_j}{\partial x_i} + \frac{\partial u_i}{\partial x_j} \right) \quad (1.18)$$

Equations 1.17 and 1.18 are the expressions of *Stokes's law of viscosity*.

Combining Eqs. 1.15 through 1.18 yields the set known as the *Navier-Stokes equations*, that is, in vector notation,

$$\rho \left(\frac{\partial \mathbf{V}}{\partial t} + \mathbf{V} \cdot \nabla \mathbf{V} \right) = \mathbf{f} - \nabla p + \mu \nabla^2 \mathbf{V} \quad (1.19)$$

Although there is no general solution of Eq. 1.19, Schlichting (1960) presents a number of solutions for particular flow situations.

When Eq. 1.19 is applied to an inviscid flow ($\mu = 0$) where the body force is the weight of the fluid acting in the negative z direction, the result is called *Euler's equation:*

$$\rho\left(\frac{\partial \mathbf{V}}{\partial t} + \mathbf{V} \cdot \nabla \mathbf{V}\right) = -g\mathbf{k} - \nabla p \qquad (1.20)$$

The nonlinear term $\mathbf{V} \cdot \nabla \mathbf{V}$ can be replaced using the vector identity

$$\mathbf{V} \cdot \nabla \mathbf{V} = \nabla\left(\frac{V^2}{2}\right) - \mathbf{V} \times (\nabla \times \mathbf{V}) \qquad (1.21)$$

In Section 1.3 it is shown that the curl of the velocity must be zero if the velocity can be represented by the potential function ϕ, that is, curl $(\mathbf{V}) = \nabla \times \mathbf{V} = 0$ when the flow is irrotational or potential. Thus Eq. 1.21 for irrotational flow is $\mathbf{V} \cdot \nabla \mathbf{V} = \nabla(V^2/2)$, and Euler's equation becomes

$$\nabla\left(\frac{\partial \phi}{\partial t} + \tfrac{1}{2}V^2 + gz\right) + \frac{\nabla p}{\rho} = 0 \qquad (1.22)$$

The scalar product of Eq. 1.22 and the directional element $d\mathbf{r} = dx\,\mathbf{i} + dy\,\mathbf{j} + dz\,\mathbf{k}$ results in the total differential of the scalar variables. This product can be integrated between any two points in the flow to obtain the following:

$$\frac{\partial \phi}{\partial t} + \tfrac{1}{2}V^2 + gz + \int \frac{dp}{\rho} = f(t) \qquad (1.23)$$

where for compressible flows the mass density is a function of the pressure. Equation 1.23 is *Bernoulli's equation*, which is a mathematical statement of the conservation of energy in an irrotational flow. When the flow is incompressible, Eq. 1.23 is simply

$$\frac{\partial \phi}{\partial t} + \tfrac{1}{2}V^2 + gz + \frac{p}{\rho} = f(t) \qquad (1.24)$$

This form of Bernoulli's equation is one of the basic equations of both linear and nonlinear wave theories found in Chapter 2.

1.5. VISCOUS FLOWS

Viscosity is that property of a fluid that causes fluid particles to adhere to an adjacent solid boundary. *Viscous* or *shear flows* normally occur in the near region of an interface of two dissimilar fluids or of a fluid and a solid. The shear flow region is called the *boundary layer*. For example, *wind-generated waves* are the result of the viscous action of the wind blowing across the air-sea interface.

The flow in the boundary layer may be *laminar*, *turbulent*, or *transitional* (*mixed*), depending on both the streamwise position and the free-stream velocity. Since the flow in the boundary layer is retarded as the fluid particles advance because of the shear action, at some streamwise position the fluid particles may lose all kinetic energy and momentarily come to rest. These particles are then swept away from the interface, and the flow is said to have *separated*. The flow region just downstream from the point of separation is called the *wake* and is characterized by eddy currents and high pressures. All of these viscous flows are represented schematically in Figure 1.6.

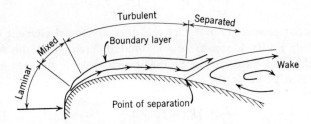

Figure 1.6. Viscous flow around a solid.

For the purposes of this book, the importance of these flows lies in the forces they induce on various bodies. Hence the rest of this discussion is devoted to flow-induced forces on the most basic of ocean engineering structures—the circular cylinder. The viscous flows encountered by this structure do not lend themselves to mathematical analyses; therefore the analyses here must be empirical in nature, that is, relying on experimental data.

Consider a circular cylinder inclined at an angle α to the horizontal direction, as in Figure 1.7. The normal and tangential velocity components, with respect to the center line of the cylinder, are

$$V_n = u \sin(\alpha) - w \cos(\alpha) \tag{1.25}$$

and

$$V_t = u \cos(\alpha) + w \sin(\alpha) \tag{1.26}$$

1.5. VISCOUS FLOWS

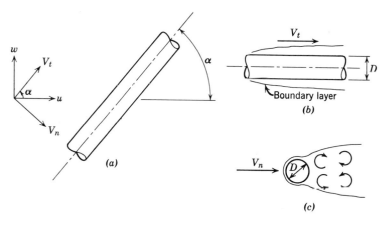

Figure 1.7. Normal and tangential velocity components on a yawed circular cylinder. (a) Geometric relations. (b) Tangential flow. (c) Normal flow.

respectively. The normal component results in both viscous and pressure forces, while the tangential component causes only a shear force tangent to the axis. These forces are collectively called the *drag*, F, and are represented in dimensionless form by the *drag coefficient, C*. The drag coefficient is found to be a function of the dimensionless coefficient called the *Reynolds number*, R_l, where l is a characteristic length parallel to the flow direction. Thus, if D is the diameter and L is the length of the cylinder, the normal and tangential drag coefficients are, respectively,

$$C_n = \frac{F_n}{\frac{1}{2}\rho V_n^2 DL} = f\left(\frac{V_n D}{\nu}\right) = f(R_D) \qquad (1.27)$$

and

$$C_t = \frac{F_t}{\frac{1}{2}\rho V_t^2 \pi DL} = f\left(\frac{V_t L}{\nu}\right) = f(R_L) \qquad (1.28)$$

These empirical relations are shown in Figure 1.8. *Note:* When the flow is laminar in the boundary layer (i.e., $R_D < 50$ and $R_L < 10^2$), then the drag is proportional to V; whereas, for a fully turbulent boundary-layer flow (i.e., $R_D > 10^2$ and $R_L > 250$ for a rough

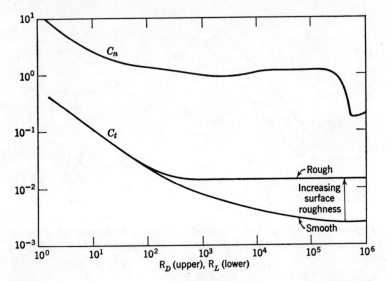

Figure 1.8. Normal and tangential drag coefficients on a yawed circular cylinder.

cylinder), the drag is proportional to V^2. The curves in Figure 1.8 are used extensively in Chapters 3 and 4 when dealing with the forces and motions on offshore structures and mooring cables.

1.6. ILLUSTRATIVE EXAMPLES

a. Collapse Depth

A steel instrumentation capsule is designed to remain in highly corrosive water for at least 2 years at a depth of 1000 ft. Because of corrosion the surface of the capsule deteriorates at a rate of 0.05 in/year. The capsule is a circular cylinder as shown in Figure 1a. The critical pressure for steel cylinders are shown in the matrix. The problem is to determine the life of the capsule, assuming the internal pressure to be 14.7 psi.

First, the hydrostatic pressure at the 1000-ft depth must be determined. Without including the effects of the compressibility of sea

1.6. ILLUSTRATIVE EXAMPLES

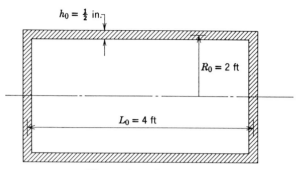

Figure 1a. Example a.

water, Eq. 1.4 yields a pressure of 444.44 psi. The inclusion of the linearly varying density results in a pressure of 444.74 psi. Therefore the compressibility can be neglected without serious error.

CRITICAL PRESSURE FOR STEEL CYLINDERS (psi)

R/L	$h/R (\times 10^2)$					
	0.4	0.8	1.2	1.6	2.0	2.4
0.3	8	45	120	260	450	720
0.4	11	64	165	350	640	990
0.5	14	78	220	450	780	1200

Next, the values of the length ratios in the pressure matrix must be determined. Initially $R_0/L_0 = 0.5000$, and $h_0/R_0 = 0.0208$. The radius-to-length ratio remains constant over the life of the capsule. Attention here is confined, therefore, to the bottom row of the matrix. After 2 years have elapsed the thickness-to-radius ratio has decreased to 0.0167. By plotting the critical pressures in the matrix as a function of h/R as in Figure 1b, one finds that the critical pressure is reached when $h/R = 0.0158$. The deterioration time for this value is

$$t = \frac{R_0(h/R) - h_0}{(\text{loss rate}) \; (h/2R - 1)} = 2.46 \text{ years}$$

Plates 1.1 and 1.2 on pages 17 and 18, respectively, show an application of a pressure hull.

16   REVIEW OF HYDROMECHANICS

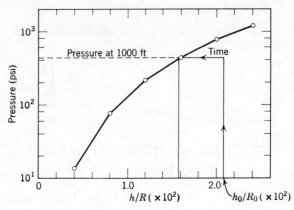

Figure 1b. Example *a*.

b. *Hydrofoil Lift*

A seismic array is towed in a body upon which two hydrofoils are mounted as shown in Figure 1c. The lift vector of the foils points away from the free surface to keep the body away from the towing vessel. The theoretical lift per unit span of the foil at any spanwise position y is given, from Valentine (1959), by an *elliptic lift distribution*, that is,

$$L' = \rho V \Gamma_0 \left[1 - \left(\frac{2y}{b} \right)^2 \right]^{1/2} \tag{1.29}$$

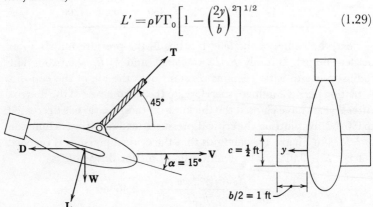

Figure 1c. Example *b*.

Plate 1.1. The pressure hull of the Deep Submergence Rescue Vehicle (DSRV) undergoing testing at the Naval Ship Research and Development Center in Carderock, Maryland. The spherical hull has been found to be the most stable when subject to severe pressures. (Courtesy of the NSRDC, U.S. Navy.)

Plate 1.2. The Deep Submergence Rescue Vehicle (DSRV) under way. This vehicle is designed to rescue crews of disabled submarines at extreme depths. (Courtesy of the NSRDC, U.S. Navy.)

1.6. ILLUSTRATIVE EXAMPLES

where Γ_0 is the circulation at the root of each foil, defined by

$$\Gamma_0 = \pi c V \sin(\alpha) \qquad (1.30)$$

The net weight (weight minus buoyancy) of the body is 10 lb, and the towing speed is 15 fps. The tension in the cable and the drag on the body are to be determined.

The lift on each foil is obtained by integrating Eq. 1.29 over the span, that is,

$$\frac{L}{2} = \int_0^{b/2} \rho V \Gamma_0 \left[1 - \left(\frac{2y}{b}\right)^2\right]^{1/2} dy = \frac{\rho V \Gamma_0 b}{8} = 144 \text{ lb}$$

so that $L = 288$ lb. Equating the force in the horizontal and vertical directions, respectively, yields the following:

$$T \cos(45°) = D + L \sin(15°)$$

and

$$T \sin(45°) = W + L \cos(15°)$$

Thus the tension is 393 lb, and the drag is 214 lb.

c. Drag on Cylinders and Reynolds Scaling

A 1-ft-diameter cylinder is placed normally to a flow of sea water ($\rho = 2.0$ slugs/ft^3 and $\nu = 1.05 \times 10^{-5}$ ft^2/sec), the velocity of which is 1 fps. The problem is to find the value of the drag per unit length of cylinder.

Also, if a model test is to be conducted in a wind tunnel ($\rho = 2.33 \times 10^{-3}$ slug/ft^3, and $\nu = 1.6 \times 10^{-4}$ ft^2/sec) on a 1-in.-diameter cylinder, the air velocity required for dynamic similarity must be determined.

The Reynolds number based on the diameter of the 1-ft cylinder is $R_D = VD/\nu = 0.953 \times 10^5$. From Figure 1.8 the drag coefficient is $C_n = 1.2$ (approximately). Therefore, when Eq. 1.27 is rearranged, the force per unit length is $F_n/L = \frac{1}{2}\rho V_n^2 D C_n = 1.2$ lb/ft. Dynamic similarity requires that the Reynolds numbers be the same for both prototype and model. Using the value of 0.953×10^5 for the 1-in. cylinder yields a velocity of 183 fps of air.

Plate 1.3. A 24-in. propeller experiencing tip cavitation (low-pressure boiling) during tests in a variable-pressure water tunnel at the Naval Ship Research and Development Center in Carderock, Maryland. (Courtesy of the NSRDC, U.S. Navy.)

1.7. SUMMARY

A knowledge of the basics of fluid mechanics is essential for one intending to work on mechanics problems in the ocean environment. If the reader does not fully understand the concepts in this chapter, he should consult the listed references to obtain the background knowledge necessary to comprehend the next chapter.

1.8. REFERENCES

King, D. A. (1969), *Handbook of Ocean and Underwater Engineering*, edited by J. J. Myers et al., McGraw-Hill, New York, Chapter 2.

Newell, H. E. (1955), *Vector Analysis*, McGraw-Hill, New York, p. 66.

Schlichting, H. (1960), *Boundary Layer Theory*, McGraw-Hill, New York.

Shames, I. H. (1962), *Mechanics of Fluids*, McGraw-Hill, New York.

Valentine, H. R. (1959), *Applied Hydrodynamics*, Butterworths, London, pp. 172–178.

1.9. PROBLEMS

1. If the deterioration rate of the capsule in Example 1.6a is 0.10 in./year, determine the life of the capsule. Derive the deterioration time equation.

2. A hydrofoil boat is designed to travel at a speed of 40 knots (1 knot = 1.689 fps) in calm water. The boat weighs 50 tons and is supported by two identical foils, one forward and one aft,

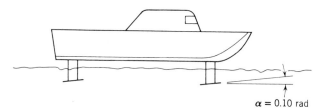

Problem 2. Hydrofoil boat.

when "flying." The angle of attack of each foil is 0.10 rad, and the chord, c, of each is 2.0 ft. Referring to Example 1.6b, determine the value of the span, b, of each foil. Assume that the boat travels in salt water, where $\rho = 2.0$ slugs/ft^3.

3. *Cavitation* is a term for the low-pressure boiling of a liquid. This phenomenon occurs when the value of the absolute pressure at a point equals that of the vapor pressure of the liquid.

Consider a fully submerged high-speed body traveling at a speed of 75 fps. The vapor pressure of the water in which it

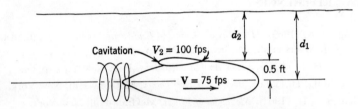

Problem 3. Cavitating high-speed body.

travels is 0.25 psi (absolute). The relative flow velocity at the point of minimum pressure on the hull is 100 fps. Determine the maximum mean depth of the body at which cavitation will occur, using Bernoulli's equation and assuming $f(t) = 0$. The diameter of the body at the minimum pressure point is 1 ft.

4. Derive the following relations in polar coordinates:

$$v_r = \frac{\partial \phi}{\partial r} = \frac{1}{r}\frac{\partial \psi}{\partial \theta}$$

$$v_\theta = \frac{1}{r}\frac{\partial \phi}{\partial \theta} = -\frac{\partial \psi}{\partial r}$$

Use Eq. 1.13 and the relations $x = r \cos(\theta)$, $z = r \sin(\theta)$, and $r = (x^2 + z^2)^{1/2}$.

5. The potential function for the flow around an infinitely long circular cylinder is

$$\phi = R^2 V_0 \left(\frac{x}{x^2 + z^2}\right) + V_0 x = RV_0 \left(\frac{R}{r} + \frac{r}{R}\right)\cos(\theta)$$

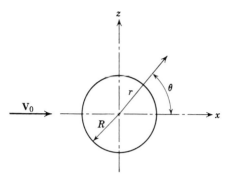

Problem 5. Flow about an infinitely long, circular cylinder.

(a) Determine the stream function expression if the flow velocity at infinity, that is, V_0, is 10 fps and the radius, R, of the cylinder is 1 ft.
(b) By letting, $\psi = 0$, plot the body profile. Let $\psi = \pm 1$ and plot the resulting streamlines.
(c) What is the velocity at $r = R$, $\theta = \pi/2$?
(d) If the center of the cylinder is 10 ft beneath the free surface, what is the minimum free-stream velocity that will cause cavitation inception at $r = R$, $\theta = \pi/2$?

6. A line array is towed horizontally at a speed of 20 fps fully submerged in deep water. The array, which is 3 in. in diameter and 100 ft in length, is enveloped in a turbulent boundary layer. Determine the viscous drag on the smooth cylinder, using the results in Figure 1.8 and assuming $\rho = 2.00$ slugs/ft^3 and $\nu = 1.05 \times 10^{-5}$ ft^2/sec. Also, let $C_t = 2.5 \times 10^{-3}$ for $R_L \geq 10^6$.

Chapter TWO

SURFACE WAVES

Waves on the free surface of a liquid exist because of nature's tendency to remain in states of equilibrium. When an object is dropped into a still pool of water, a disturbance is created in the form of a *surface wave*. The subsequent motions of the surface are the result of the gravitational action tending to return the water to its undisturbed position. Since these waves result from gravitational attraction, they are also referred to as *gravity waves*.

An excellent physical description of surface waves is contained in the book by Bascom (1964). Mathematical treatments of the subject can be found in the publications by Kinsman (1965) and Phillips (1966) and in the classical treatise of Lamb (1945).

In this chapter the reader is introduced to both the mathematical and physical descriptions of surface waves. The linear wave theory and two nonlinear theories—those of Stokes (1847, 1880) and of Gerstner (1809)—are discussed.

2.1. LINEAR WAVE-THEORY

All waves, whether gravity, acoustic, or electromagnetic, obey some form of the *wave equation*. The dependent variable in each case

2.1. LINEAR WAVE THEORY

depends on the physical phenomenon, as do the boundary conditions. In general, the wave equation and the boundary conditions may be either linear or nonlinear. The incompressible flow in a surface wave must satisfy the special linear form of the wave equation known as *Laplace's equation*, if the flow is assumed to be irrotational. In this section the solution of Laplace's equation is obtained and is subjected to linearized boundary conditions.

Before embarking on the analysis of linear waves, it will be helpful to consider the physical characteristics of a *traveling surface wave*, as shown schematically in Figure 2.1. The origin of the Cartesian coordinate system is located on the *still-water level* (*SWL*), which is the

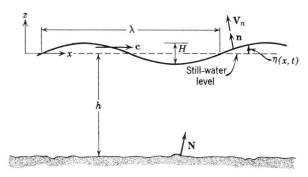

Figure 2.1. Schematic drawing of a traveling surface wave.

undisturbed position of the water. The *depth*, h, of the water is measured from the sea floor to the SWL. The wave itself has a *height*, H, measured from the trough to the crest, a *wavelength*, λ, and a *velocity*, c, which is sometimes referred to as the *celerity*. The *displacement* of the free surface from the SWL at any position x and time t is denoted by η.

The *boundary conditions* that must be satisfied by the water within the wave are the following:

1. No water particles can cross the free surface. In other words, particles on the free surface must always remain there. To satisfy

this condition the particle velocity at $z = \eta$ must be equal to the normal velocity of the free surface, that is,

$$\mathbf{V}|_{z=\eta} = \mathbf{V}_n \tag{2.1a}$$

This condition is referred to as the *kinematic surface condition*. If the flow is assumed to be irrotational, then the fluid velocity can be represented by the *potential function*, ϕ, as in Eq. 1.10. Therefore Eq. 2.1a can be written as

$$\mathbf{V}|_{z=\eta} = \nabla \phi|_{z=\eta} = \frac{\partial \phi}{\partial n}\bigg|_{z=\eta} \mathbf{n} \tag{2.1b}$$

where $\mathbf{n}$ is the *outward unit normal vector* on the free surface. Furthermore, it is assumed that η is very small compared to the wavelength, so that the condition described by Eq. 2.1 approximately occurs at $z = 0$. This approximation makes it possible to replace the *normal coordinate*, n, by z and the normal unit vector, $\mathbf{n}$, by $\mathbf{k}$. Thus Eq. 2.1b becomes

$$\mathbf{V}\bigg|_{z=\eta} \simeq \frac{\partial \eta}{\partial t} \mathbf{k} = \frac{\partial \phi}{\partial z}\bigg|_{z=0} \mathbf{k} \tag{2.2}$$

2. On the sea floor the fluid particles cannot cross the solid boundary. Mathematically, at $z = -h$

$$\mathbf{V} \cdot \mathbf{N} = \frac{\partial \phi}{\partial N} = 0 \tag{2.3}$$

where N is the normal coordinate at the sea floor, and $\mathbf{N}$ is the outward unit normal vector.

3. The pressure on the free surface is zero (gauge) for any position x and any time t. If the flow is assumed to be irrotational, Bernoulli's equation (Eq. 1.24) applied at the free surface $(z = \eta)$ is

$$\frac{\partial \phi}{\partial t} + g\eta + \tfrac{1}{2}V^2 = 0 \tag{2.4}$$

where the free surface is taken as the datum so that the time function in Eq. 1.24 is zero. Equation 2.4, called the *dynamic surface condition*, is nonlinear because of the velocity-squared term. For small dis-

2.1. LINEAR WAVE THEORY 27

placements of the free surface the nonlinear term, being of second order in comparison to the other two terms, can be neglected. The *linearized dynamic condition* is then

$$\eta = -\frac{1}{g}\frac{\partial \phi}{\partial t}\bigg|_{z=\eta} \tag{2.5}$$

Physically, this linearization assumes that the flow is slow enough to result in a kinetic energy of the fluid particles that is much less than the other mechanical energies.

A question normally asked concerning the linearizing of the free-surface conditions is, "How small is a small displacement?" The answer to this question can be obtained by considering the surface motions of a specific wave. Physical observations of deep-water waves show that the particles on the surface travel in nearly circular paths, the radii of which are equal to the wave amplitude, $H/2$, as shown in Figure 2.2. If a wave passes a particle over a time period T

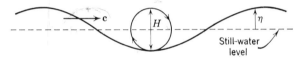

Figure 2.2. Approximate particle motion on the free surface.

(which is equal to the inverse of the wave frequency), the particle travels a distance πH with an average velocity of $\pi H/T$.

Using this information, consider two waves having lengths of 100 ft and periods of 4.5 sec, but respective heights of 10 ft and 1 ft. The particles on the surface of the steeper wave ($H = 10$ ft) have a speed of 7.85 fps, while those of the shallow wave travel at 0.785 fps. Comparing the last two terms of Eq. 2.4, applied at a crest ($\eta = H/2$), yields values for the larger wave of $gH/2 = 161$ ft^2/sec^2 and $\frac{1}{2}V^2 = 30.8$ ft^2/sec^2. The corresponding values for the smaller wave are $gH/2 = 16.1$ ft^2/sec^2 and $\frac{1}{2}V^2 = 0.308$ ft^2/sec^2. The difference in these values for the larger wave is only half an order of magnitude; however, the difference for the smaller wave is one and one-half orders. In general, we can assume that the linearized equations apply for values of H/λ up to $\frac{1}{50}$.

The linearized free-surface conditions of Eqs. 2.2 and 2.5 are now combined by eliminating η to obtain

$$\left[\frac{1}{g}\frac{\partial^2 \phi}{\partial t^2} + \frac{\partial \phi}{\partial z}\right]_{z=\eta \simeq 0} = 0 \qquad (2.6)$$

We are dealing with a continuous fluid within the wave; therefore the equation of continuity must be satisfied. Again, if the flow is assumed to be irrotational, the velocity may be represented by the potential function, as in Eq. 1.10, allowing the continuity equation to be written as

$$\nabla^2 \phi = 0 \qquad (2.7)$$

which is Laplace's equation.

The analysis from this point consists simply of obtaining a general solution of Eq. 2.7 and requiring the solution to satisfy the boundary conditions expressed by Eqs. 2.3 and 2.6. Since Eq. 2.7 is an *elliptic partial differential equation*, a product solution is assumed of the form

$$\phi = X(x) Z(z) T(t) \qquad (2.8)$$

Substituting this expression into Eq. 2.7, dividing the result by XZT, and then separating the variables yields

$$\frac{1}{X}\frac{d^2 X}{dx^2} = -\frac{1}{Z}\frac{d^2 Z}{dz^2} = -k^2 \qquad (2.9)$$

where k is a constant called the *wave number*. It should be mentioned here that the sign of k^2 in the separated equation is determined from a knowledge of the physical phenomenon. It is known that the wave is periodic in the x direction; therefore the sign of k^2 must be such that the x-indicial equation is complex.

The general solutions for the z and x equations are, respectively,

$$Z(z) = C_1 \cosh(kz + \alpha) \qquad (2.10)$$

and

$$X(x) = C_2 \sin(kx + \beta) \qquad (2.11)$$

where C_1, C_2, α, and β are arbitrary constants. The particular form of $Z(z)$ is determined by applying the condition on the sea floor; that is, Eq. 2.3 is satisfied at $z = -h$ if

2.1. LINEAR WAVE THEORY

$$\frac{dZ}{dz} = C_1 k \sinh(-kh + \alpha) = 0$$

from which $\alpha = kh$. Thus Eq. 2.10 becomes

$$Z(z) = C_1 \cosh(kz + kh) \tag{2.12}$$

The linearized free-surface condition in Eq. 2.6 can be used to determine the time-dependent function of Eq. 2.8. Substitution of Eqs. 2.8 and 2.12 into Eq. 2.6 yields

$$\left[\frac{1}{g}\cosh(kz+kh)\frac{d^2T}{dt^2} + kT\sinh(kz+kh)\right]\Bigg|_{z=0} = 0$$

or, upon separation of variables,

$$\frac{1}{T}\frac{d^2T}{dt^2} + \omega^2 = 0 \tag{2.13}$$

where

$$\omega = [kg \tanh(kh)]^{1/2} \tag{2.14}$$

is the *circular frequency* of the wave. The general solution of Eq. 2.13 is

$$T(t) = C_3 \sin(\omega t + \gamma) \tag{2.15}$$

The *phase angles* β and γ in Eqs. 2.11 and 2.15 can be eliminated by simply shifting the respective spatial and temporal origins.

By combining Eqs. 2.11, 2.12, and 2.15 with Eq. 2.8 and letting $\beta = \gamma = 0$, the following expression for the velocity potential is obtained:

$$\phi = A \cosh(kz + kh) \sin(kx) \sin(\omega t) \tag{2.16}$$

where the coefficients C_1, C_2, and C_3 have been absorbed in the *amplitude coefficient*, A. Using the expression for the velocity potential in the dynamic free-surface condition, Eq. 2.5, yields the expression for the displacement of the free surface of a *standing wave*:

$$\eta = -\frac{1}{g}\frac{\partial \phi}{\partial t}\bigg|_{z=0} = -\frac{\omega A}{g}\cosh(kh)\sin(kx)\cos(\omega t)$$

$$= a \sin(kx) \cos(\omega t) \tag{2.17}$$

where a is the *wave amplitude*. The amplitude coefficient A in Eq. 2.16 can be replaced by $-ag/\omega \cosh(kh)$ to obtain

$$\phi = -\frac{ag}{\omega} \frac{\cosh(kz+kh)}{\cosh(kh)} \sin(kx) \sin(\omega t) \qquad (2.18)$$

In addition to the wave described by Eq. 2.17, there are three additional standing wave solutions that are combinations of the sine and cosine functions of x and t. Thus the four standing waves are described by

$$\eta_1 = a \sin(kx) \cos(\omega t)$$
$$\eta_2 = a \cos(kx) \cos(\omega t)$$
$$\eta_3 = a \sin(kx) \sin(\omega t)$$

and

$$\eta_4 = a \cos(kx) \sin(\omega t)$$

These expressions result from the solution of a linear equation; therefore the property of *superposition* allows the addition of any pair of them to obtain additional solutions. These new solutions, such as

$$\eta^- = \eta_1 + \eta_4 = a \sin(kx + \omega t) \qquad (2.19)$$

and

$$\eta^+ = \eta_2 + \eta_3 = a \cos(kx - \omega t) \qquad (2.20)$$

describe *traveling waves*, where the superscripts plus and minus refer to *left-running* and *right-running waves*, respectively. A right-running wave is shown schematically in Figure 2.1. The velocity potentials corresponding to Eqs. 2.19 and 2.20, respectively, are

$$\phi^- = \frac{ag}{\omega} \frac{\cosh(kz+kh)}{\cosh(kh)} \cos(kx + \omega t) \qquad (2.21)$$

and

$$\phi^+ = \frac{ag}{\omega} \frac{\cosh(kz+kh)}{\cosh(kh)} \sin(kx - \omega t) \qquad (2.22)$$

Attention from this point is confined here to right-running waves. Consider the wave described by Eq. 2.20. If the origin of the coordinate system is allowed to travel with the wave, then the angle of the cosine term in the equation is constant, that is,

$$kx - \omega t = \text{constant}$$

The differential of the angle is, then,

$$k\,dx - \omega\,dt = 0$$

from which the wave velocity or celerity is

$$c = \frac{dx}{dt} = \frac{\omega}{k} = \frac{\lambda}{T} = f\lambda \qquad (2.23)$$

where f is the wave frequency in hertz, λ is the wavelength, and T is the period. Combining Eqs. 2.14 and 2.23 yields the following:

$$c = \left[\frac{g}{k}\tanh(kh)\right]^{1/2} \qquad (2.24)$$

and

$$\lambda = cT = \frac{gT^2}{2\pi}\tanh\left(\frac{2\pi h}{\lambda}\right) \qquad (2.25)$$

which is a *transcendental equation* for the wavelength. An example of a numerical solution of Eq. 2.25 is given in Example 2.6c. The wavelength expression shows that λ approaches $gT^2/2\pi$ as the depth approaches infinity, however, as the water becomes shallow, the wavelength approaches $(gh)^{1/2}T$. In other words, the wavelength (and celerity) is a strong function of the period in deep water and a function of both the depth and the period in shallow water. It is also evident that the wavelength (and celerity) decreases with depth in shallow water.

Now consider the particle motion. The *velocity components* of the fluid in a right-running wave are obtained by combining Eqs. 1.10 and 2.22. The results are as follows:

$$u = \frac{\partial \phi}{\partial x} = \frac{agk}{\omega}\frac{\cosh(kz+kh)}{\cosh(kh)}\cos(kx-\omega t) \qquad (2.26)$$

and

$$w = \frac{\partial \phi}{\partial z} = \frac{agk}{\omega}\frac{\sinh(kz+kh)}{\cosh(kh)}\sin(kx-\omega t) \qquad (2.27)$$

32 SURFACE WAVES

Consider a fixed point (x_0, z_0) in the wave. From Eqs. 2.26 and 2.27 it is evident that u and v vary sinusoidally in time at that position, indicating that the fluid particles have (x_0, z_0) as a *mean point*. Thus, in Figure 2.3, two new coordinates, ξ and ζ, can be introduced,

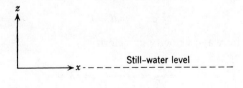

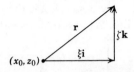

Figure 2.3. Position vector of a fluid particle.

with their origins at (x_0, z_0), in order to define the instantaneous position of a fluid particle. The velocity components of the particle can be expressed in terms of the new coordinates as

$$u = \frac{d\xi}{dt} \tag{2.28}$$

and

$$w = \frac{d\zeta}{dt} \tag{2.29}$$

Equation 2.26 is combined with 2.28 and Eq. 2.27 with 2.29 by eliminating u and v, respectively, and the results are integrated from $t = 0$ to $t = t$ to obtain the expressions for the displacement components ξ and ζ:

$$\xi = \int_0^t \frac{\partial \phi}{\partial x} dt \bigg|_{x_0, z_0} = -\frac{agk}{\omega^2} \frac{\cosh(kz_0 + kh)}{\cosh(kh)} \sin(kx_0 - \omega t) \tag{2.30}$$

and

$$\zeta = \int_0^t \frac{\partial \phi}{\partial z} dt \bigg|_{x_0, z_0} = \frac{agk}{\omega^2} \frac{\sinh(kz_0 + kh)}{\cosh(kh)} \cos(kx_0 - \omega t) \tag{2.31}$$

2.1. LINEAR WAVE THEORY

Equations 2.30 and 2.31 show that, for any mean position (x_0, z_0), the paths of the fluid particles are determined by the ratio of the depth to the wavelength, $h/\lambda = kh/2\pi$. To illustrate consider the particle *position vector* **r**, as shown in Figure 2.3. Using the expressions in Eqs. 2.30 and 2.31 with the relationships in Eqs. 2.23 and 2.24 yields the position vector

$$\mathbf{r} = \xi\mathbf{i} + \zeta\mathbf{k}$$

$$= \frac{a}{\sinh(kh)}[-\cosh(kz_0 + kh)\sin(kx_0 - \omega t)\mathbf{i} + \sinh(kz_0 + kh)\cos(kx_0 - \omega t)\mathbf{k}] \quad (2.32)$$

The values of kh in the hyperbolic functions determine the shapes of the particle paths as follows.

CASE 1. DEEP WATER $(\frac{1}{2} \leq h/\lambda < \infty)$

For this range of h/λ values, $\sinh(kh) \simeq \cosh(kh) \simeq e^{kh}/2$ and $\tanh(kh) \simeq 1$. Equation 2.32, therefore, is approximately

$$\mathbf{r} = ae^{kz_0}[-\sin(kx_0 - \omega t)\mathbf{i} + \cos(kx_0 - \omega t)\mathbf{k}] \quad (2.33)$$

Since z_0 is negative beneath the still-water level, Eq. 2.33 describes a circular path the radius of which decreases exponentially with depth, as in Figure 2.4a. The wave in this case is called a *short wave* since its length is much less than the depth.

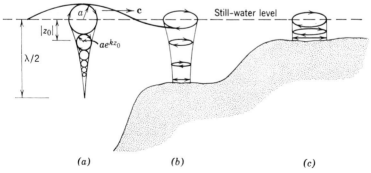

Figure 2.4. Depth effects on particle paths. (*a*) Deep water. (*b*) Water of intermediate depth. (*c*) Shallow water.

CASE 2. INTERMEDIATE DEPTH $\frac{1}{20} < h/\lambda < \frac{1}{2}$)

In this case Eq. 2.32 describes an elliptic path with major and minor axes that decrease exponentially with depth (see Figure 2.4b).

CASE 3. SHALLOW WATER ($0 < h/\lambda \leq \frac{1}{20}$)

The hyperbolic functions in this case are approximated as follows: $\sinh(kh) \simeq \tanh(kh) \simeq kh$, and $\cosh(kh) \simeq 1$. With these approximations Eq. 2.32 becomes

$$r = \frac{a}{kh}\left[-\sin(kx_0 - \omega t)\mathbf{i} + (kh + kz_0)\cos(kx_0 - \omega t)\mathbf{k}\right] \quad (2.34)$$

which describes an elliptic path. The difference between this case and the intermediate one is that the major axis of the ellipse is independent of depth in shallow water, as illustrated in Figure 2.4c. The waves in shallow water are sometimes referred to as *long waves* since the length is much greater than the depth.

It is of interest to study the wave properties in the extreme cases of shallow and deep water. First, consider the celerity and length of a wave in shallow water. When the shallow-water approximations are applied to Eqs. 2.24 and 2.25, respectively, the celerity and wavelength expressions are

$$c = (gh)^{1/2} \quad (2.35)$$

and

$$\lambda = \frac{2\pi}{k} = (gT^2h)^{1/2} \quad (2.36)$$

From these relations it is evident that both the celerity and the wavelength decrease with depth in shallow water.

Now consider the horizontal velocity component of the water particles in shallow water. Equation 2.26 yields

$$u = \frac{agk}{\omega}\cos(kx - \omega t) = \frac{ag}{(gh)^{1/2}}\cos(kx - \omega t) \quad (2.37)$$

Thus the particle velocity increases with decreasing depth. Com-

2.1. LINEAR WAVE THEORY   35

paring Eqs. 2.35 and 2.37, one sees that there is a depth at which the velocity component, u, and the celerity, c, are equal. When this equality occurs at a wave crest, that is, when $\cos(kx - \omega t) = 1$, the wave is said to *break*. When u exceeds c, the wave is said to *spill*. The shape of the wave at the break predicted by the linear wave theory is sinusoidal; however, experience shows that a breaking wave has a pointed crest, as illustrated in Figure 2.5. This inability to describe the actual wave profile is one of the deficiencies of the linear wave theory. Returning to the mathematics, one finds, by equating Eqs. 2.35 and 2.37, using the results of Eq. 2.36, that the depth of water at the break is $h = a$.

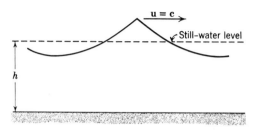

Figure 2.5. Schematic profile of a breaking wave.

Physically, the breaking condition is one of *instability*. The fluid in the wave is constantly adjusting its velocity to satisfy the conservations of both mass and energy. When these conditions can no longer be simultaneously satisfied, the wave breaks and then spills, dissipating the energy of the water into turbulence.

Can breaking occur in deep water? Obviously the mathematical condition of the equality of u at a crest and c can be obtained. Applying the deep-water approximations to Eqs. 2.24, 2.25, and 2.26 and equating u and c at a crest, one obtains $a = 1/k$. Physically, breaking occurs in deep water when an energy source external to the fluid sufficiently accelerates the surface water. For example, on a windy day *white caps* (waves spilling after breaking) can be observed in deep water. The shear action of the wind on the free surface increases the particle velocity, u, so that breaking occurs.

The dynamical aspects of breaking waves are discussed in Chapter 3.

2.2. THE WAVE GROUP

While looking out on a rather calm sea one might observe a patch of waves traveling in a given direction. These waves could be generated by a gust of wind or by a school of fish traveling near the free surface. The waves within the patch are said to be traveling in a *group*.

To study the behavior of the waves in the group, begin by considering two linear waves of equal amplitude but slightly differing lengths. Referring to Eq. 2.20, and assuming the property of superposition, one can describe the displacement of the free surface by

$$\eta = a\{\cos(kx - \omega t) + \cos[(k + \Delta k)x - (\omega + \Delta\omega)t]\} \quad (2.38a)$$

or, using trigonometric identities,

$$\eta = 2a \cos\left(\frac{\Delta k}{2} x - \frac{\Delta \omega}{2} t\right) \cos\left[\tfrac{1}{2}(2k + \Delta k)x - \tfrac{1}{2}(2\omega + \Delta\omega)t\right] \quad (2.38b)$$

In Eq. 2.38b the variation of the first cosine function is much less than that of the second since $k \gg \Delta k$ and $\omega \gg \Delta \omega$. Hence the surface profile is similar to that shown schematically in Figure 2.6. If the angle of the first cosine is held constant, the *group velocity* is found to be

$$c_g = \frac{dx}{dt} = \lim_{\Delta k \to 0} \frac{\Delta \omega}{\Delta k} = \frac{d\omega}{dk} \quad (2.39)$$

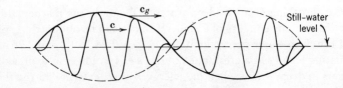

Figure 2.6. The wave group.

Similarly, if the angle of the second cosine function of Eq. 2.38b is assumed to be constant, the velocity of the individual waves within the group is

$$c = \lim_{\Delta k \to 0} \frac{\omega + \Delta\omega/2}{k + \Delta k/2} = \frac{\omega}{k}$$

$$= \left[\left(\frac{g}{k}\right) \tanh (kh)\right]^{1/2} \tag{2.40}$$

where the result of Eq. 2.24 has been used.

The relation between the group velocity and the celerity is then

$$c_g = \frac{d\omega}{dk} = \frac{d(kc)}{dk}$$

$$= \tfrac{1}{2}c\left[1 + \frac{2kh}{\sinh (2kh)}\right] \tag{2.41}$$

Thus in deep water $kh \to \infty$ and $c_g \to c/2$, whereas in shallow water $kh \to 0$ and $c_g \to c$. Physically, in deep water waves appear at the rear of the group and travel to the front, where they disappear. This phenomenon can be observed in a deep-water wave tank. A wave group is created by turning the wave generator on for a few cycles and then turning it off.

The concept of the wave group is important in determining the wave drag on fixed structures and displacement vessels, topics discussed in Chapter 3.

2.3. WAVE ENERGY

A surface wave is the result of energy being transmitted to the water from an external source such as a ship, an earthquake, or a gust of wind. The wave itself can be considered as the transmitting agent of this energy away from its source. In this section the expressions for the kinetic and potential energies are derived and also the equation for the energy flux in a linear wave.

Consider the wave shown in Figure 2.7. If a unit depth into the page is assumed, the mass of the element of water above the still-water level is

$$\Delta m = \rho\eta \, (\Delta x)$$

SURFACE WAVES

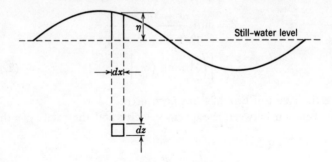

Figure 2.7. A linear wave.

and its potential energy is

$$\frac{\Delta m \, g\eta}{2} = \frac{\rho g \, \Delta x \, \eta^2}{2}$$

Replacing the displacement term, η, by the expression in Eq. 2.20 and integrating the result over the wavelength yields the expression for the *potential energy*:

$$E_p = \frac{\rho g a^2}{2} \int_0^\lambda \cos^2 (kx - \omega t) \, dx$$

$$= \frac{\rho g a^2}{2k} \left[\frac{\xi}{2} + \frac{\sin (\xi) \cos (\xi)}{2} \right] \Bigg|_{-\omega t}^{k\lambda - \omega t}$$

$$= \frac{\rho g a^2 \lambda}{4} \qquad (2.42)$$

where ξ is an integration variable.

Now, consider a submerged element of mass of water $\rho \, \Delta x \, \Delta z$, as shown in Figure 2.7. The kinetic energy of this mass is

$$\Delta E_k = \tfrac{1}{2} \rho (u^2 + w^2) \, \Delta x \, \Delta z$$

where, for a linear wave, u and w are represented by the expressions in Eqs. 2.26 and 2.27, respectively. The total *kinetic energy* over one wavelength is

2.3. WAVE ENERGY

$$E_k \simeq \tfrac{1}{2}\rho \int_{-h}^{0}\int_{0}^{\lambda}(u^2+w^2)\,dx\,dz$$

$$= \frac{\tfrac{1}{2}\rho a^2 g^2 k^2}{\omega^2 \cosh^2(kh)} \int_{-h}^{0}\int_{0}^{\lambda} [\cosh^2(kz+kh)\cos^2(kx-\omega t) +$$

$$\sinh^2(kz+kh)\sin^2(kx-\omega t)]\,dx\,dz$$

$$= \frac{\rho g a^2 \lambda}{4} \tag{2.43}$$

where the upper limit of the z integration is taken to be zero since the displacement of the free surface at any horizontal position is assumed to be very small, and where

$$\int {\cosh^2(a\xi) \atop \sinh^2(a\xi)}\,d\xi = \frac{1}{4a}\sinh(2a\xi) \pm \frac{\xi}{2} \tag{2.44}$$

and, from Eq. 2.14,

$$\omega^2 = gk \tanh(kh) \tag{2.45}$$

At this point the reader may question the use of the term (u^2+w^2) in the kinetic energy expression since it is assumed to be of second order in the linear wave derivation, that is, in the derivation of the dynamic free-surface condition, Eq. 2.5. The answer is that the term $\tfrac{1}{2}V^2$ is assumed to be negligible when compared to the other energy forms *at a point* in the water, that is, Bernoulli's equation is a point relation; however, when integrated over the entire wave, the total kinetic energy is not negligible. As expected, the total potential and kinetic energies are of equal magnitude.

As previously mentioned, the energy per unit volume at a point is expressed by Bernoulli's equation (Eq. 1.24) in the following form:

$$\rho \frac{\partial \phi}{\partial t} = -\rho g z - p + F(t) \tag{2.46}$$

To obtain the expression for the *energy flux*, that is, the time rate of change of energy per unit area normal to the flow direction, one simply multiplies Eq. 2.46 by the particle velocity:

$$\Delta \dot{E} = \rho \frac{\partial \phi}{\partial t}\mathbf{V} = \rho \frac{\partial \phi}{\partial t}\nabla \phi$$

By replacing the velocity potential, ϕ, by the expression in Eq. 2.22 and integrating the result over the wave period, T, and water depth, h, the expression for the *average energy flux per wave* is found to be

$$\dot{\mathbf{E}} \simeq -\frac{\rho a^2 g^2}{Tc \cosh^2(kh)} \int_{-h}^{0} \int_{0}^{T} [\cosh(kz+kh)\cos(kx-\omega t) \cdot$$
$$\cosh(kz+kh)\cos(kx-\omega t)\mathbf{i} + \sinh(kz+kh) \cdot$$
$$\sin(kx-\omega t)\mathbf{k}] \, dt \, dz$$

$$= \frac{\rho g a^2 c}{4}\left[\frac{2kh}{\sinh(2kh)} + 1\right]\mathbf{i}$$

$$= \frac{\rho g a^2 c_g}{2}\mathbf{i} \qquad (2.47)$$

where referring to Eq. 2.41, c_g is the group velocity. The integral of the $\mathbf{k}$ term of Eq. 2.47 is naturally equal to zero since the velocity component w is parallel to the vertical plane.

If the *total energy expression*, $E = E_p + E_k$, is divided by the wavelength, the resulting expression is that of the average energy per wave:

$$\bar{E} = \frac{E}{\lambda} = \tfrac{1}{2}\rho g a^2$$

Comparison of this expression and Eq. 2.47 shows that the average energy flux in the direction of wave travel is simply the product of the average energy and the group velocity. From the results of Section 2.2, it is evident that in deep water the energy is transmitted at the half-wave velocity; in shallow water, however, the energy transmission occurs at the celerity.

2.4. NONLINEAR WAVES

It was stated in Section 2.1 that the linear theory cannot be used to predict the profile of a breaking wave. This statement can be generalized to the prediction of the profiles of waves of *finite height*, that is, waves for which the height-to-length ratio, H/λ, is greater than $\frac{1}{50}$. Waves of finite height are, therefore, nonlinear in nature. It will be recalled that the linearization in the wave theory of Section 2.1 occurs in the derivation of the dynamic free-surface condition, where the V^2 term of Eq. 2.4 is assumed to be of second order. For waves of finite height this assumption cannot be made.

There are several nonlinear wave theories, including the perturbation theory of Stokes (1847, 1880) and the geometric theory of Gerstner (1809). The two theories mentioned are discussed in this section for the sake of completeness. The use of these theories in the analysis of engineering problems is not practical, however, since both are rather cumbersome. The linear theory has actually been proved to be the most useful in the analysis of engineering situations.

A. Stokes's Theory

Stokes in his 1880 publication developed a higher-order method of analysis of gravity waves of finite height. Basically, *Stokes's theory* consists of assuming that the properties of wave motion, such as the velocity potential, can be represented by a series of small *perturbations*. Therefore the higher the number of terms included in the series, the better is the approximation of the actual wave properties. Rather complete reviews of Stokes's perturbation technique can be found in the publications of Kinsman (1965) and of Neumann and Pierson (1966).

Consider the right-running wave of finite amplitude illustrated in Figure 2.8, where the origin of the coordinate system travels with the crest of the wave. As in the linear analysis of Section 2.1, Laplace's equation (Eq. 2.7) and the boundary conditions at the free surface

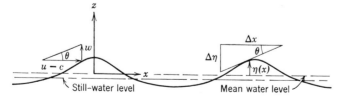

Figure 2.8. Profile of a wave of finite height.

42 SURFACE WAVES

and on the sea floor must be satisfied. If the flow is assumed to be irrotational, Bernoulli's equation applied to the quasi-steady flow at the surface is

$$g\eta + \tfrac{1}{2}[(u-c)^2 + w^2] = K \text{ (constant)} \qquad (2.48)$$

Referring to Figure 2.8, one can obtain the kinematic free-surface condition from geometric consideration since the slope of the free surface at any point is

$$\tan\theta = \lim_{\Delta x \to 0} \frac{\Delta \eta}{\Delta x} = \frac{d\eta}{dx} = \frac{w}{u-c} \qquad (2.49)$$

The vertical component of the particle velocity on the surface is then

$$w = \frac{\partial \phi}{\partial z} = \frac{d\eta}{dx}\left(\frac{\partial \phi}{\partial x} - c\right), \qquad z = \eta \qquad (2.50)$$

The velocity potential, ϕ, must now be determined. Since the free surface is always in the neighborhood of the mean-water level (i.e., $z=0$), the partial derivatives of the potential function on the free surface can be represented by a *Maclaurin series*, that is, at $z=\eta$

$$\frac{\partial \phi}{\partial x} = \sum_{m=0}^{\infty} \frac{1}{m!} \left.\frac{\partial^{m+1}\phi}{\partial x \, \partial z^m}\right|_{z=0} \eta^m \qquad (2.51)$$

and

$$\frac{\partial \phi}{\partial z} = \sum_{m=0}^{\infty} \frac{1}{m!} \left.\frac{\partial^{m+1}\phi}{\partial z^{m+1}}\right|_{z=0} \eta^m \qquad (2.52)$$

The free-surface condition as expressed by Eqs. 2.48 and 2.50, respectively, can now be written as

$$g\eta + \frac{1}{2}\left(\sum_{m=0}^{\infty} \frac{1}{m!} \left.\frac{\partial^{m+1}\phi}{\partial x \, \partial z^m}\right|_{z=0} \eta^m - c\right)^2 + \frac{1}{2}\left(\sum_{m=0}^{\infty} \frac{1}{m!} \left.\frac{\partial^{m+1}\phi}{\partial z^{m+1}}\right|_{z=0} \eta^m\right)^2 = K \qquad (2.53)$$

and

$$\sum_{m=0}^{\infty} \frac{1}{m!} \left.\frac{\partial^{m+1}\phi}{\partial z^{m+1}}\right|_{z=0} \eta^m = \frac{d\eta}{dx}\left(\sum_{m=0}^{\infty} \frac{1}{m!} \left.\frac{\partial^{m+1}\phi}{\partial x \, \partial z^m}\right|_{z=0} \eta^m - c\right) \qquad (2.54)$$

2.4. NONLINEAR WAVES

Now, let the velocity potential, free-surface displacement, and celerity, respectively be represented in perturbation form as follows:

$$\phi(x, z) = \sum_{m=1}^{n} \alpha^m \phi_m \qquad (2.55)$$

$$\eta(x) = \sum_{m=1}^{n} \alpha^m \eta_m \qquad (2.56)$$

and

$$c = \sum_{m=0}^{n} \alpha^m c_m \qquad (2.57)$$

where the coefficients of the powers of the small number α are independent of α. Thus, to satisfy Laplace's equation (Eq. 2.7), the seafloor condition of Eq. 2.3, and the free-surface conditions of Eqs. 2.53 and 2.54, the coefficients of the powers of α in these equations must be identically zero. The equations formed by these coefficients comprise the *n*th-*order theories* of Stokes as follows.

From Eq. 2.53, after letting $K = \sum_{m=0}^{n} \alpha^m K_m$,

α^0:
$$\tfrac{1}{2} c_0^2 = K_0 \qquad (2.58)$$

α^1:
$$g\eta_1 - \frac{1}{2}\left(c_0 \frac{\partial \phi_1}{\partial x} - c_0 c_1\right)\bigg|_{z=0} = K_1 \qquad (2.59)$$

α^2:
$$g\eta_2 + \frac{1}{2}\left[\left(\frac{\partial \phi_1}{\partial x}\right)^2 - 2c_0 \frac{\partial \phi_2}{\partial x} - 2c_1 \frac{\partial \phi_1}{\partial x} - 2c_0 \frac{\partial^2 \phi_1}{\partial x \, \partial z} \eta_1 + c_1^2 + 2c_0 c_1 + \left(\frac{\partial \phi_2}{\partial z}\right)^2\right]\bigg|_{z=0} = K_2 \qquad (2.60)$$

etc.

The coefficients of α^n in Eq. 2.54 are

α^0:
$$0 = 0 \qquad (2.61)$$

α^1:
$$\frac{\partial \phi_1}{\partial z}\bigg|_{z=0} = -c_0 \frac{d\eta_1}{dx} \qquad (2.62)$$

α^2:
$$\left(\frac{\partial \phi_2}{\partial z} + \frac{\partial^2 \phi_1}{\partial z^2}\eta_1\right)\bigg|_{z=0} = \frac{d\eta_1}{dx}\frac{\partial \phi_1}{\partial x}\bigg|_{z=0} - c_1 \frac{d\eta_1}{dx} - c_0 \frac{d\eta_2}{dx} \qquad (2.63)$$

etc.

44   SURFACE WAVES

Laplace's equation yields the following:

$$n: \quad \frac{\partial^2 \phi_n}{\partial x^2} + \frac{\partial^2 \phi_n}{\partial z^2} = 0 \qquad (2.64)$$

and, finally, the sea-floor condition is

$$n: \quad \nabla \phi_n \cdot \mathbf{k} = 0 \qquad (2.65)$$

at $z = -h$.

To illustrate the method, consider the *first-order theory*, that is, $n = 1$. As in the linear analysis, Laplace's equation, Eq. 2.64 with $n = 1$, and the boundary condition on the sea floor, Eq. 2.65, must be satisfied by ϕ_1. Thus the equations for the first-order theory are the following:

Laplace's equation:

$$\nabla^2 \phi_1 = 0 \qquad (2.66)$$

Sea-floor condition:

$$\left. \frac{\partial \phi_1}{\partial z} \right|_{z=-h} = 0 \qquad (2.67)$$

Dynamic free-surface condition:

$$g\eta_1 - c_0 \left(\frac{\partial \phi_1}{\partial x} - c_1 \right) \bigg|_{z=0} - K_1 = 0 \qquad (2.68)$$

Kinematic free-surface condition:

$$\left. \frac{\partial \phi_1}{\partial z} \right|_{z=0} + c_0 \frac{d\eta_1}{dx} = 0 \qquad (2.69)$$

The solution of Eq. 2.66 which satisfies Eq. 2.67 is

$$\phi_1 = A_1 \cosh (kh + kz) \sin (kx) \qquad (2.70)$$

where A_1 is a constant. A derivative of Eq. 2.68 with respect to x is taken, and the result is combined with 2.69 by eliminating $d\eta_1/dx$. Thus, at $z = 0$,

$$\frac{-g}{c_0} \frac{\partial \phi_1}{\partial z} - c_0 \frac{\partial^2 \phi_1}{\partial x^2} = 0$$

2.4. NONLINEAR WAVES

Combining this relation with the velocity potential expression of Eq. 2.70 yields

$$c_0 = \left[\frac{g}{k} \tanh(kh)\right]^{1/2} \quad (2.71)$$

The displacement of the free-surface is obtained by integrating Eq. 2.69 with respect to x while using the relation in 2.70:

$$\eta_1 = -\int \frac{1}{c_0} \frac{\partial \phi_1}{\partial z}\bigg|_{z=0} dx = \frac{A_1}{c_0} \sinh(kh) \cos(kx) \quad (2.72a)$$

or, using the results of Eq. 2.71, one obtains

$$\eta_1 = \frac{k A_1 c_0}{g} \cosh(kh) \cos(kx) \quad (2.72b)$$

When the expressions for ϕ_1 and η_1 are substituted into Eq. 2.68, the result is

$$c_0 c_1 = K_1 \quad (2.73)$$

Returning to the series representations of Eqs. 2.55 and 2.56, one writes the first-order approximations of the velocity potential and the surface displacement as

$$\phi = A_1 \cosh(kh + kz) \sin(kx) \quad (2.74)$$

and

$$\eta = \alpha \frac{A_1 c_0 k}{g} \cosh(kh) \cos(kx)$$

$$= \frac{H}{2} \cos(kx) \quad (2.75)$$

Note: Instead of the amplitude, a, the wave height, H, is used in these equations since the profile of the nonlinear wave is not symmetric about the still-water level.

By comparing the last two terms of Eq. 2.75, the perturbation constant is found to be

$$\alpha = \frac{H}{2} \frac{g}{c_0 A_1 k \cosh(kh)} \quad (2.76)$$

SURFACE WAVES

Since time is not present in Eqs. 2.74 and 2.75, these expressions are those of a fixed wave. The corresponding velocity potential and free-surface displacement for a traveling wave are obtained by fixing the coordinate system in space and assuming that ϕ and η are functions of time, as was done in Section 2.1. The resulting expressions are

$$\phi = \frac{H}{2} \frac{g}{kc_0} \frac{\cosh(kh+kz)}{\cosh(kh)} \sin(kx - \omega t) \tag{2.77}$$

and

$$\eta = \frac{H}{2} \cos(kx - \omega t) \tag{2.78}$$

which are identical with the results of the linear theory as given by Eqs. 2.22 and 2.20, respectively.

Without derivation, the velocity potential and the free-surface displacement expressions obtained from the *second-order theory* of Stokes are, respectively,

$$\phi = \frac{H}{2} c \frac{\cosh(kh+kz)}{\sinh(kh)} \sin(kx - \omega t) + \frac{H^2}{4} \frac{3\pi c}{4\lambda} \frac{\cosh(2kh+2kz)}{\sinh^4(kh)} \sin(2kx - 2\omega t) \tag{2.79}$$

and

$$\eta = \frac{H}{2} \cos(kx - \omega t) + \frac{H^2}{4} \frac{\pi}{2\lambda} \frac{\cosh(kh)}{\sinh^3(kh)} \cdot$$
$$[2 + \cosh(2kh)] \cos(2kx - 2\omega t) \tag{2.80}$$

where the celerity expression is

$$c = \left[\frac{g}{k} \tanh(kh)\right]^{1/2} = c_0 \tag{2.81}$$

The second-order theory yields reasonably good results when the depth-to-wavelength ratio is greater than $\frac{1}{10}$, which is a practical range for most engineering applications. In addition to approximating the profile of the free surface, the second-order theory predicts the shape of a breaking wave. The angle of the breaking crest obtained from the theory is 120°, as shown in Figure 2.9. Finally, the

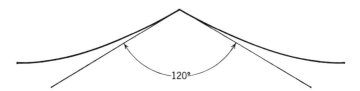

Figure 2.9. Breaking wave profile as predicted by Stokes.

theory predicts a net horizontal *convection* of the fluid particles per wave cycle, a phenomenon that has also been physically observed. The convection velocity in deep water as derived by Stokes is

$$U_{\text{con}} = \frac{H^2}{4} k^2 c e^{2kz_0} \tag{2.82}$$

The convection, therefore, is maximum on the free-surface and diminishes exponentially with depth, recalling that z is positively directed upward. The path of the particles about any mean depth z_0 is similar to that shown schematically in Figure 2.10. The treatment of the theory given by Wehausen and Laitone (1960) includes the complete mathematical details of these findings.

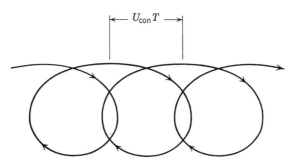

Figure 2.10. Convection path beneath a finite wave.

B. Gerstner's "Trochoidal" Theory

The nonlinear rotational wave theory of Gerstner (1809) is referred to as the "*trochoidal*" *theory* because of the predicted profiles of the constant-pressure surfaces resulting from it, that is, these profiles are generated by points on a wheel traveling on a staight line located above the x axis. The radius of the wheel must be equal to $1/k$ since one revolution of the wheel occurs over a distance equal to one wavelength. The wave height, H, is then determined by the radial location of the generating point. In the case of Figure 2.11, if the generating point is on the circumference, the curve generated is that of a

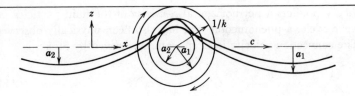

Figure 2.11. Generation of the trochoidal wave profile.

cycloid with a *cusp* at the crest (in Figure 2.9, the angle at the crest is zero). The shape is thus that of a breaking wave of height $2/k = \lambda/\pi$. As the generating point approaches the center of the wheel, the shape of the constant-pressure surface approaches that of the linear theory, that is, sinusoidal.

The equations of the trochoidal theory can be derived from those of the linear theory of Section 2.1 as follows. Consider the components of the deep-water particle paths described by Eq. 2.33:

$$\xi = -\frac{H}{2} e^{kz_0} \sin(kx_0 - \omega t) = x - x_0$$

and

$$\zeta = \frac{H}{2} e^{kz_0} \cos(kx_0 - \omega t) = z - z_0$$

where (x_0, z_0) is the position of the center of the generating wheel. Thus a constant-pressure surface is described by

2.4. NONLINEAR WAVES

$$x = x_0 - \frac{H}{2} e^{kz_0} \sin(kx_0 - \omega t) \tag{2.83}$$

and

$$z = z_0 + \frac{H}{2} e^{kz_0} \cos(kx_0 - \omega t) \tag{2.84}$$

These equations describe a trochoid generated by a wheel traveling to the left of the page, as shown schematically in Figure 2.11. The wave, however, travels to the right at a speed of

$$c = \left(\frac{g}{k}\right)^{1/2} \tag{2.85}$$

from the linear theory.

As derived by Kinsman (1965), the velocity components of the fluid particles are

$$u = \frac{dx}{dt} = \frac{H}{2} \omega e^{kz_0} \cos(kx_0 - \omega t) = \omega(z - z_0) \tag{2.86}$$

and

$$w = \frac{dz}{dt} = \frac{H}{2} \omega e^{kz_0} \sin(kx_0 - \omega t) = -\omega(x - x_0) \tag{2.87}$$

From these expressions it is obvious that the particles travel in circular orbits, just like the paths predicted by the linear theory. The radii of the orbits, again, decrease exponentially with depth, as illustrated in Figure 2.12. The velocity components of Eqs. 2.86 and 2.87 satisfy the equation of continuity as expressed by Eq. 1.8; however, these components do not satisfy the condition of irrotationality, namely, $\nabla \times \mathbf{V} = 0$, since

$$\nabla \times \mathbf{V} = \frac{\partial u}{\partial z} - \frac{\partial w}{\partial x} = \omega\left(2 - \frac{\partial z_0}{\partial z} - \frac{\partial x_0}{\partial x}\right)$$

$$= \Omega \tag{2.88}$$

where Ω is the *vorticity*. Thus the flow described by the trochoidal theory is rotational with a vorticity value, from Kinsman (1965), of

$$\Omega = -\frac{\omega H^2 k^2 e^{2kz_0}}{2 - H^2 k^2 e^{2kz_0}/2} \neq 0 \tag{2.89}$$

50 SURFACE WAVES

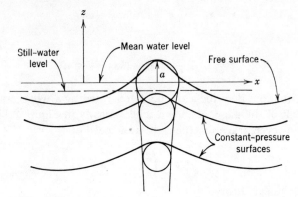

Figure 2.12. Constant-pressure surfaces and particle orbits beneath a trochoidal wave.

Hence the vorticity decreases with depth. The vortical direction is negative and, therefore, in the opposite direction of the particle motions. This is one of the inadequacies of Gerstner's theory.

The trochoidal theory is presently used primarily by naval architects in analyses of ship structures. These are, however, static analyses since the theory is applied only to obtain a wave profile that approximates the largest waves encountered by a ship. Although the theory is useful in this type of engineering analysis, it does not lend itself readily to the dynamic situations encountered by ocean engineers.

2.5. EXPERIMENTAL WAVE STUDY

The value of any theory lies in its ability to accurately predict actual physical phenomena. Therefore this section, which contains descriptions of a wave-measuring device and of an actual wave study, is presented as a supplement to the theoretical study of waves.

a. Wave Measurement

There are a number of wave-measuring devices, some of which are practical only for laboratory studies. These include the *digital wave staff* described by Hedges (1967), *capacitance probes*, *floats*, and the *resistance wire gauge*, to name a few examples.

2.5. EXPERIMENTAL WAVE STUDY

A device that is easily constructed and somewhat easy to use in laboratory studies is the resistance wire gauge [Compton (1965)]. This gauge, shown schematically in Figure 2.13, consists of two parallel wires separated by a small distance, d. The wires are connected to a Wheatstone bridge as shown in the figure. While the gauge is out of water, there is an infinite resistance between the wires. When the probe is immersed, the water acts as a conductor, and the resistance in the probe circuit depends on the length of the wires above the water.

Three problems are associated with this type of probe. First, the physical separation of the wires places a lower limit on the size of the wave being studied; that is, waves that have lengths of the order of magnitude of d cannot be accurately measured. Second, caution must be taken when using two probes simultaneously, since each probe can influence the signal of the other when the gauges are separated by a short distance. Finally, the output of the probe is nonlinear.

Although these problems exist, the resistance wire gauge has been found to be most satisfactory in the laboratory. An idea of the size and use of the gauge can be obtained from the following description of an actual experiment.

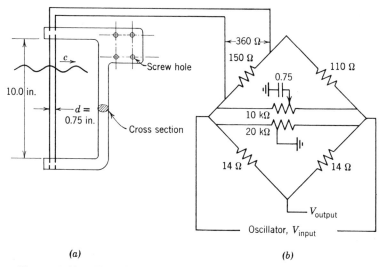

Figure 2.13. The resistance wire wave gauge. (*a*) Gauge. (*b*) Bridge.

b. Wave Experiment

To gain confidence in the various wave theories, an experimental study of wave properties is conducted in a wave tank using two resistance wire gauges. The tank is 4 ft deep, and the generated waves can have a maximum height of 3 in. and a maximum length of 8 ft. Signals from the gauges are recorded on a two-channel strip-chart recorder. The system is shown schematically in Figure 2.14.

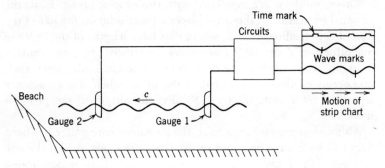

Figure 2.14. Schematic diagram of a wave experiment.

First, the gauges must be calibrated by raising and lowering each one from its measuring position and simultaneously recording the signals. During this static calibration the gauges are found to interfere with each other at a separation distance of 2 ft. No interference is encountered, however, when the gauges are separated by 6 ft.

Since the gauges are separated by a distance that is greater than the lengths of the waves being studied, specific wave crests should be noted on the chart as they pass each gauge, as in Figure 2.14. This allows the investigator to compute the celerity or phase velocity by simply measuring the time interval during which the crest travels over the gauge separation length.

The frequency of the waves is obtained from the recorded signal of either gauge by counting the number of crests that pass over a given time period.

Results of such an investigation are shown in Figure 2.15, along with the wave profile calculated by using the linear theory, that is,

2.5. EXPERIMENTAL WAVE STUDY

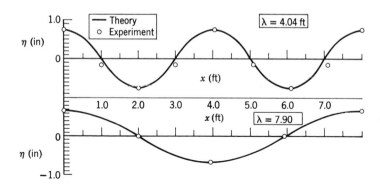

Figure 2.15. Experimental and theoretical wave profiles.

Eq. 2.20, since this theory is employed extensively in the following chapters. Note that for the steeper wave (i.e., $H/\lambda = 1/32.3$) the agreement between the experimental data and the theoretical curve is good. As the amplitude of the wave increases, however, the actual wave profile becomes more pointed at the crest and broader in the trough.

The experimental and theoretical celerity and wavelength values are given in the table, the theoretical values being determined from Eqs. 2.24 and 2.25 applied to deep water.

Wave Value		Experimental		Theoretical	
f (Hz)	H/λ	c (fps)	λ (ft)	c (fps)	λ (ft)
1.143	1/32.3	4.62	4.04	4.56	3.99
0.800	1/70.2	6.32	7.90	6.37	7.97

The agreement between the experimental and theoretical values is good. Hence one can now use the linear theory with some confidence for small values of H/λ.

2.6. ILLUSTRATIVE EXAMPLES

a. Circular Surface Waves

Waves can be created on the surface of a still pool of water by dropping a small object to create a disturbance. The resulting wave pattern is similar to that shown in Figure 2a. If the flow is assumed to be irrotational, the condition of continuity is given by Laplace's equation (Eq. 2.7). In this case, however, the equation must be written in circular cylindrical coordinates since the pattern is symmetric about a vertical axis through the point at which the dropped object hits:

$$\nabla^2 \phi = \frac{\partial^2 \phi}{\partial r^2} + \frac{1}{r}\frac{\partial \phi}{\partial r} + \frac{1}{r^2}\frac{\partial^2 \phi}{\partial \beta^2} + \frac{\partial^2 \phi}{\partial z^2} = 0 \qquad (2.90)$$

The solution of Eq. 2.90 is obtained in the same manner as that for Cartesian coordinates, that is, by assuming a product solution. Thus one assumes

$$\phi = R(r)B(\beta)Z(z)T(t) \qquad (2.91)$$

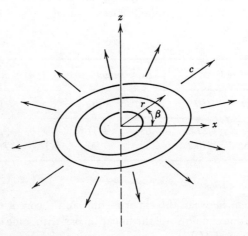

Figure 2a. Circular surface wave pattern.

2.6. ILLUSTRATIVE EXAMPLES

Combining Eqs. 2.91 and 2.90 and dividing the result by the product $RBZT$ results in the following:

$$\frac{1}{R}\frac{d^2R}{dr^2} + \frac{1}{Rr}\frac{dR}{dr} + \frac{1}{Br^2}\frac{d^2B}{d\beta^2} + \frac{1}{Z}\frac{d^2Z}{dz^2} = 0$$

Since r, β, and z are independent variables, this equation can be rearranged as follows:

$$-\frac{1}{R}\left(R'' + \frac{R'}{r}\right) - \frac{B''}{Br^2} = \frac{Z''}{Z} = k^2$$

where the primes and double primes indicate first and second differentiation, respectively, with respect to the variables associated with the dependent variable in question. The constant k is, again, a wave number. Replacing the z function by k^2, as indicated in the equation, yields the following:

$$-\frac{r^2}{R}\left(R'' + \frac{R'}{r}\right) - k^2 r^2 = \frac{B''}{B} = -m^2$$

where m is a constant.

The radial equation is then

$$R'' + \frac{R'}{r} + \left(k^2 - \frac{m^2}{r^2}\right)R = 0 \qquad (2.92)$$

which is called *Bessel's equation*. A rather complete coverage of these and other special equations can be found in, for example, Churchill (1963).

The solution of this equation is written in terms of the *Bessel function of order m*, that is,

$$R = C_r J_m(kr) \qquad (2.93)$$

where the Bessel function is defined by

$$J_m(kr) = \sum_{j=0}^{\infty} \frac{(-1)^j}{j!(m+j)!}\left(\frac{kr}{2}\right)^{m+2j}, \qquad m = 0, 1, 2, \ldots \quad (2.94)$$

Actually, there are two kinds of Bessel functions that satisfy Eq. 2.92. These are called the Bessel function of the *first kind*, denoted as $J_m(kr)$, and that of the *second kind*, denoted as $Y_m(kr)$ or sometimes

$N_m(kr)$ since it is also referred to as the *Neumann function*. Again, the text by Churchill (1963) provides complete details. The function $J_m(kr)$ has a wave-like curve for any value m. In particular, the zeroth-order function, that is, $m = 0$, is shown in Figure 2b.

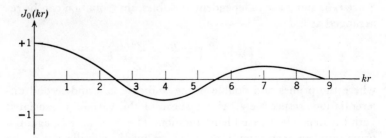

Figure 2b. Bessel function of the first kind, zeroth order.

The solutions of the β and z equations are obtained in the same manner as that used in the solutions of the x and z equations of Section 2.1. Thus

$$B = C_\beta \sin(m\beta + \alpha)$$

and

$$Z = C_z \cosh(kh + kz)$$

The general solution for the velocity potential representing the flow beneath a circular wave is then (with no β-variation)

$$\phi = A J_0(kr) \cosh(kh + kz) f(t) \tag{2.95}$$

The important difference between this type of wave and the two-dimensional linear type is that the energy per unit vertical area diminishes as the circular wave travels from its origin, whereas the energy in the linear wave remains constant. The reason for this energy intensity decrease is that the wave front of the circular wave expands as the wave travels away from its source; therefore, since the total energy in a wave is constant, the intensity must decrease as the wave front increases its circumferential dimension.

2.6. ILLUSTRATIVE EXAMPLES

b. Interfacial Waves

A layer of oil of thickness h_0 and density ρ_0 lies on a layer of water of thickness h_w. The linear wave analysis can be used to determine the properties of the waves at the fluid interface.

Assuming h_0 is large enough to prevent the free surface of the oil from being disturbed, one applies Bernoulli's equation to each fluid at the interface, that is (referring to Figure 2c), at $z = \eta$. Thus, since the displacement of each fluid from the still-water position is η, Eq. 1.24 yields

$$\rho_w \frac{\partial \phi_w}{\partial t} + \rho_w g \eta + p = \text{constant}$$

and

$$\rho_0 \frac{\partial \phi_0}{\partial t} + \rho_0 g \eta + p = \text{constant}$$

where the velocity-squared term is neglected in the linearization.

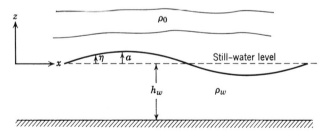

Figure 2c. Linear wave at the oil-water interface.

Note that the pressure and the interfacial displacement are the same for both fluids since the oil and water must remain in contact. The pressure and the constant terms can be eliminated in combining the Bernoulli equations to obtain the following:

$$\rho_w \frac{\partial \phi_w}{\partial t} - \rho_0 \frac{\partial \phi_0}{\partial t} + g\eta(\rho_w - \rho_0) = 0 \qquad (a)$$

The difference in the velocity potentials is only in the depth behavior, that is, the function described in Eq. 2.12. Applying the

58 SURFACE WAVES

condition of no vertical flow at $z = +h_0 \to \infty$ and at $z = -h_w$ results in

$$Z_0 = \lim_{h_0 \to \infty} [C_0' \cosh(kz - kh_0)] = C_0 e^{-kz}$$

and

$$Z_w = C_w \cosh(kz + kh_w)$$

from Eq. 2.12. The potential functions are now

$$\phi_0 = C_0 e^{-kz} \sin(kx - \omega t) \tag{b}$$

and

$$\phi_w = C_w \cosh(kz + kh_w) \sin(kx - \omega t) \tag{c}$$

The displacement of the oil-water interface from the static equilibrium position is

$$\eta = a \cos(kx - \omega t) \tag{d}$$

The kinematic condition applied at the interface requires that

$$\frac{\partial \eta}{\partial t} = \frac{\partial \phi_w}{\partial z} = \frac{\partial \phi_0}{\partial z}, \qquad z = 0 \tag{e}$$

which, when combined with Eqs. b, c, and d, results in the relations

$$a\omega = kC_w \sinh(kh_w) = -kC_0 \tag{f}$$

Using the relationships in Eqs. b, c, d, and f in the dynamic condition of Eq. a results in the following:

$$\frac{-\rho_w \omega^2}{k} \coth(kh_w) - \frac{\rho_0 \omega^2}{k} + g(\rho_w - \rho_0) = 0 \tag{g}$$

Rearranging Eq. g, one obtains the expression for the circular frequency of the wave:

$$\omega = \left[\frac{kg(\rho_w - \rho_0)}{\rho_w \coth(kh_w) + \rho_0} \right]^{1/2} \tag{2.96}$$

The corresponding expression for the celerity is

$$c = \frac{\omega}{k} = \left[\frac{g(\rho_w - \rho_0)}{k\rho_w \coth(kh_w) + k\rho_0} \right]^{1/2} \tag{2.97}$$

The primary assumption in this example is that the layer of oil is of sufficient thickness to ensure that no motions occur on the free surface of the oil. This assumption requires that $h_0 > \lambda/2$ so that $\cosh(kh_0) \simeq 1$.

c. A Computer Solution of Wavelength by Successive Approximations

Equation 2.25 describes the wavelength as

$$\lambda = \frac{gT^2}{2\pi} \tanh\left(\frac{2\pi h}{\lambda}\right)$$

where T and h are assumed to be experimentally determined. This equation is transcendental and, therefore, must be solved numerically.

To obtain the solution, one begins by defining a difference function by

$$\Delta(\lambda) = \frac{gT^2}{2\pi} \tanh\left(\frac{2\pi h}{\lambda}\right) - \lambda = F(\lambda) - G(\lambda)$$

where the functions $F(\lambda)$ and $G(\lambda)$ are plotted in Figure 2d. The desired value of λ then must be a root of the difference function $\Delta(\lambda)$. This root is determined by the method of *successive approximations*, as described in the text by McCracken and Dorn (1964).

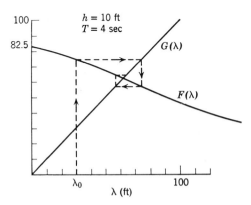

Figure 2d. Wavelength relations solved by successive approximations.

First, choose a value of the wavelength, say $\lambda_0 = 1$ in. Use this value in the difference expression to calculate $\Delta(\lambda_0)$. Compare the value of the difference function with some acceptable upper limit, for example, $\Delta(\lambda) = 0.001$. If

$$|\Delta(\lambda) - 0.001| > 0$$

then replace λ_0 by $F(\lambda_0)$ to obtain λ_1. Repeat this procedure until $|\Delta(\lambda_n) - 0.001| = 0$ to a desired accuracy. Thus λ_n is the solution of Eq. 2.25.

The method of successive approximations is easily performed on a digital computer. Before writing the program, a flow diagram is constructed as in Figure 2e. From this diagram, the following FORTRAN IV program is developed:

```
       READ 10, PI, G
       READ 10, T, H
       PRINT 10, T, H
       X = 1.0
   1   Y = 2.*PI*H/X
       TANHY = (EXP(Y) - EXP(-Y))/(EXP(Y) + EXP(-Y)
       Z = (G*T**2/(2.*PI))*TANHY
       IF(ABS(Z - X) - 0.001)5, 5, 3
   3   X = Z
       GØ TØ 1
   5   PRINT 20, Z
  10   FØRMAT(E12.4, E12.4)
  20   FØRMAT(E12.5)
```

where PI $= \pi$, G $= g$, T $= T$, H $= h$, and X $= \lambda$.

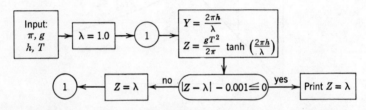

Figure 2e. Flow diagram of computer solution of wavelength.

2.7. SUMMARY

This chapter is devoted primarily to the description and derivation of the kinematic properties of surface waves. The dynamic properties are discussed in the next chapter, which deals with wave forces on fixed structures.

The linear wave theory is used extensively in the following chapters since, as the reader has observed, it is both reasonably accurate and relatively simple. The vibratory motions discussed in Chapter 3 and the heaving and pitching motions covered in Chapter 4 are assumed to be linear. Thus the use of the linear wave as an excitation of these motions is appropriate in the analyses.

2.8. REFERENCES

Bascom, W. (1964), *Waves and Beaches*, Doubleday, New York.

Churchill, R. V. (1963), *Fourier Series and Boundary Value Problems*, 2nd ed., McGraw-Hill, New York.

Compton, R. H. (1965), "The Resistance Wire Gauge," *Proceedings of 14th General Meeting of American Towing Tank Conference*, Webb, Institute of Naval Architecture, Glen Cove, N.Y.

Gerstner, F. J. (1809), "Theorie der Wellen," *Ann. Physik*, **32,** 412–440.

Hedges, C. P. (1967), "Digital Wave Staff," *Transactions of 3rd Annual Conference of Marine Technology Society*.

Kinsman, B. (1965), *Wind Waves*, Prentice-Hall, Englewood Cliffs, N.J.

Lamb, H. (1945), *Hydrodynamics*, Dover Publications, New York.

McCracken, D., and Dorn, W. (1964), *Numerical Methods and Fortran Programming*, John Wiley, New York.

Neumann, G., and Pierson, W. (1966), *Principles of Physical Oceanography*, Prentice-Hall, Englewood Cliffs, N.J.

Phillips, O. M. (1966), *The Dynamics of the Upper Ocean*, Cambridge University Press, London.

Stokes, G. G. (1847), "On the Theory of Oscillatory Waves," *Trans. Cambridge Phil. Soc.*, **8**, and *Papers*, **I**, 197.

——— (1880), Supplement to the 1847 paper under the same title, *Cambridge Phil. Soc. Math. Phys. Papers*. **I**, 314.

Wehausen, J., and Laitone, E. (1960), "Surface Waves," in *Handbuch der Physik*, Springer-Verlag, Berlin, pp. 446–778.

2.9. PROBLEMS

1. A deep-water wave having a length of 100 ft and a height of 5 ft travels toward shore.
   (a) What are the values of the length and celerity at the position where the water is 2 ft in depth?
   (b) What are the values of the total wave energy and the energy flux (per unit width) of the wave in deep water? (Assume salt water so that $\rho = 2.00$ slugs/ft^3.)

2. Determine the expression for the stream function of a standing linear wave by using the velocity potential of Eq. 2.18 and the Cauchy-Riemann relations of Eq. 1.13. Plot the streamline corresponding to $\psi = 0$. What does this streamline represent?

3. Deep-water waves that are 25 ft long and 1 ft high travel into a cold-water region where the mass density of the water is $\rho = 2.10$ slugs/ft^3. The value of ρ in the original warm-water region is 1.90 slugs/ft^3. If the values of the wave frequency and length are unchanged, what is the wave height in the cold water?

4. Wind blows across a deep-water wave that is 10 ft in length. If the wind accelerates the particles on the surface sufficiently to cause the wave to break, what is the height of the breaking wave?

5. Wind blows across waves having a height of 0.5 ft. Determine the following:
   (a) the length of the breaking wave in deep water;
   (b) the depth at which the wave breaks in shallow water.

6. Determine the expression for the height of a wave breaking in intermediate water.

7. A group of waves is 1000 ft in length in deep water. The waves within the group are 10 ft in length.
   (a) Calculate the time taken for a component wave to travel the length of the group.
   (b) How far will the group have progressed during this time?

2.9. PROBLEMS

8. Write a FORTRAN IV program to determine the dimensionless wave profile, using the results of Stokes's theory in Eq. 2.80 and assuming $t = 0$; that is, determine the relation of η/H and x/λ.
9. A small float is released from a deep ocean structure and subsequently travels 500 ft in 1 hr while in 5-ft waves. From this information determine the length of the waves.
10. Using the results of Problem 8, determine the profile of a deep-water wave 100 ft long and 5 ft high. Compare this profile with that obtained from the linear theory.
11. What is the value of the vorticity for the surface particles of the waves described in Problem 10?
12. Using the expression for the velocity potential in Eq. 2.95 and replacing r in that equation by x, determine the expression for the displacement of the free surface, assuming $f(t) = \sin(\omega t)$.
13. The reader has probably heard the statement "Oil calms troubled waters." From the results presented in Example 2.6b can the validity of this expression be established? *Hint:* Consider the total energy of the waves in deep water.
14. Using the method described in Example 2.6c, determine the length of the deep-water wave when it travels into 20 ft of water.
15. Compare the experimental results presented in Figure 2.15 with those obtained using Stokes's theory and the trochoidal theory.

Chapter THREE

FIXED STRUCTURES IN WAVES

Engineers are not as concerned with the wave itself as they are with the consequences of the existence of waves. One such possible consequence is the destruction of a concrete sea wall, caused by the wave-induced pressure gradient at the wall.

In this chapter the momentum and energy transfer mechanisms between surface waves and fixed bodies are discussed. Again, the emphasis is not on specialized structures but on the structures that are universally used in engineering situations, such as the circular pile.

3.1. HYDROSTATIC PRESSURE BENEATH A SURFACE WAVE

The variation of the hydrostatic pressure within a static fluid was derived in Chapter 1 and expressed in Eq. 1.1. For moderate depths the specific weight of the fluid in Eq. 1.1 is constant for all practical purposes; therefore the expression for the pressure variation can be integrated directly to obtain the pressure at any depth, that is,

$$p = \int_z^0 \gamma \, dz = -\gamma z \tag{3.1}$$

3.1. HYDROSTATIC PRESSURE BENEATH A SURFACE WAVE

where z has its origin on the still-water-level and is directed positively upward.

If waves are present on the free surface, then Eqs. 2.26 and 2.27 show that there is significant fluid motion to a depth of approximately one-half a wavelength in deep water, or to the depth of the sea floor in intermediate or shallow water. Thus, for an irrotational flow, the pressure variation is obtained from Bernoulli's equation (Eq. 1.24), which is applied to the flow within a right-running wave, using the velocity potential expression given in Eq. 2.22. If the flow is assumed to be slow enough that the nonlinear term, that is, $\frac{1}{2}\rho V^2$, can be neglected, Bernoulli's equation is

$$p = -\rho \frac{\partial \phi}{\partial t} - \rho g z$$

$$= \rho a g \frac{\cosh (kh + kz)}{\cosh (kh)} \cos (kx - \omega t) - \rho g z \quad (3.2)$$

where the free surface is chosen to be the energy datum.

Consider the case in which $\cos (kx - \omega t) = 0$, that is, at the nodal points of the wave. In this case Eq. 3.2 reduces to Eq. 3.1, the hydrostatic equation, although the motion in the vertical direction is maximum at these points. This means that the time variation of the pressure is due to the particle motion in the horizontal direction.

Now, when $\cos (kx - \omega t) = \pm 1$, Eq. 3.2 describes the pressure beneath a crest $(+)$ or a trough $(-)$. Note, however, that the pressure does not go to zero on the free surface. Applying Eq. 3.2 at a crest or trough yields, for the *pressure "residue,"*

$$\Delta_p = \rho g \{\pm a [\cosh (ka) + \tanh (kh) \sinh (ka)] - a\} \neq 0 \quad (3.3)$$

For values of a and λ in the linear wave range, the value of the residue is negligible. This fact will be illustrated. The reason for the existence of this pressure residue on the free surface is contained in the boundary conditions of the linear wave theory. The kinematic free-surface condition, Eq. 2.2, is applied at $z = 0$; however, the dynamic free-surface condition is applied at $z = \eta$. Thus these boundary conditions are inconsistent, the inconsistency causing inaccurate prediction of the pressure at the air-sea interface.

66 FIXED STRUCTURES IN WAVES

To gain some idea of the magnitude of the pressure residue, consider deep-water waves where $h > \lambda/2$ and for which $\cosh(kh) \simeq e^{kh}/2$. For this case Eq. 3.2 is approximated by

$$p = \rho a g e^{kz} \cos(kx - \omega t) - \rho g z \qquad (3.4)$$

If $z < -\lambda/2$, then the pressure variation with depth is approximately linear, that is, the hydrostatic pressure variation is described by Eq. 3.1 beneath the region of wave influence. Now, dividing Eq. 3.4 by $\rho g \lambda$ yields

$$\frac{p}{\rho g \lambda} = \frac{a}{\lambda} e^{kz} \cos(kx - \omega t) - \frac{z}{\lambda} \qquad (3.5a)$$

or, letting $P = p/\rho g \lambda$ and $Z = z/\lambda$,

$$P = \frac{a}{\lambda} e^{2\pi Z} \cos(kx - \omega t) - Z \qquad (3.5b)$$

The results obtained from Eq. 3.5b are presented in Figure 3.1, assuming a wave steepness of $H/\lambda = \frac{1}{50}$. It can be seen that the dimensionless pressure residue at the crest and trough are rather

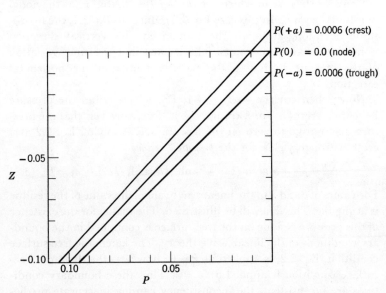

Figure 3.1. Dimensionless pressure distribution beneath a traveling wave (linear) of steepness $H/\lambda = \frac{1}{50}$.

3.2. WAVES AT A VERTICAL FLAT BARRIER

small, both having a value of 0.0006. At a depth of half of the wavelength the values of P beneath a crest, trough, and node are, respectively, 0.5004, 0.4996, and 0.500. At this depth, therefore, the values of the dimensionless pressures beneath a crest and trough are within .1% of the pressure beneath the node. Because of the negligible value of the residue pressure, the linear theory can be used with a good degree of confidence in engineering analyses.

3.2. WAVES AT A VERTICAL FLAT BARRIER

When a traveling wave strikes a vertical flat barrier such as a sea wall, the wave is reflected and a standing wave results. If the wave breaks in the vicinity of the barrier, the forces on the structure depend on whether the wave breaks well before, just before, or at the barrier. This was experimentally determined by Bagnold (1939). In this section, an analysis is presented of nonbreaking waves that are perfectly reflected at a vertical barrier.

Perfect reflection can be represented mathematically by adding the expression for the free-surface displacement of a right-running wave and that for its *mirror image*, which is a left-running wave with the same profile but a phase velocity directed to the left (see Figure 3.2). If the right-running wave is described by

$$\eta^+ = a \cos(kx - \omega t) \tag{3.6}$$

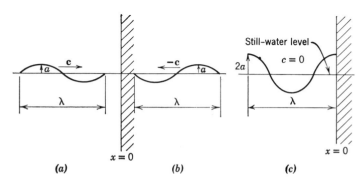

Figure 3.2. Perfect reflection of a traveling wave. (*a*) Right-running wave. (*b*) Mirror image. (*c*) Standing wave.

and its mirror image by
$$\eta^- = a \cos(kx + \omega t) \tag{3.7}$$
then the reflected wave is described by
$$\eta = \eta^+ + \eta^- = 2a \cos(kx) \cos(\omega t) \tag{3.8}$$
where the double-angle trigonometric identity has been used. Equation 3.8 describes a standing wave with an amplitude twice that of the incident traveling wave. Actually, because of viscosity and other nonideal phenomena such as elasticity and permeability of the sea wall, the standing wave amplitude is not quite twice the incident wave amplitude since these phenomena absorb part of the energy of the wave.

The larger amplitude of a standing wave can cause problems in waters where ships are moored. The motions caused by these steeper waves can result in severe stresses within the structural members of a vessel, as well as in the mooring lines. This is one of the reasons that ships are sometimes taken out to sea during storms.

The velocity potential of the standing wave is obtained by adding the potentials of the right- and left-running waves corresponding to the free-surface displacements of Eqs. 3.6 and 3.7:
$$\phi = \phi^+ + \phi^- = \frac{-(H/2)g}{\omega} \frac{\cosh(kh + kz)}{\cosh(kh)} \cos(kx) \sin(\omega t) \tag{3.9}$$
When this relation is used in Bernoulli's equation, the pressure at any depth on the sea wall is described by the following:
$$p = -\rho \frac{\partial \phi}{\partial t} - \rho g z = \frac{H}{2} \rho g \frac{\cosh(kh + kz)}{\cosh(kh)} \cos(\omega t) - \rho g z \tag{3.10}$$
where the sea wall is located at $x = 0$.

For the case in Figure 3.3, the wave-induced force on the sea wall is obtained by integrating the pressure expression of Eq. 3.10 over the wetted surface. Thus, if a unit width into the page is assumed, the *force* on the sea wall is
$$F(t) = \int_{-h}^{\eta_0(t)} p \, dz = \frac{(H/2)\rho g}{k} \frac{\sinh(kh + k\eta_0)}{\cosh(kh)} \cos(\omega t) - \frac{\rho g}{2}(\eta_0^2 - h^2)$$
$$= -\frac{\rho \omega w_0}{k^2} \cot(\omega t) - \frac{\rho g}{2}(\eta_0^2 - h^2) \tag{3.11}$$

3.2. WAVES AT A VERTICAL FLAT BARRIER

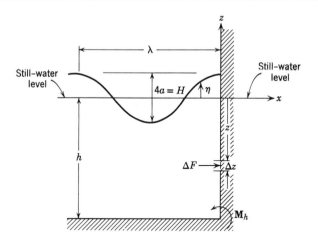

Figure 3.3. Forces and moments induced by a standing wave on a vertical flat barrier.

where η_0 is the free-surface displacement at $x = 0$, and w_0 is the vertical velocity of the fluid particle on the free surface adjacent to the barrier. The reason for introducing w_0 into the force expression is to show that the dynamic contribution to the horizontal force, F, is due to the tangential velocity of the water at the wall.

The *moment about the base* of the vertical barrier due to the time-dependent force is

$$M_h(t)^{+)} = -\int_{-h}^{\eta_0(t)} (z+h) p \, dz = -\int_{-h}^{\eta_0(t)} zp \, dz - Fh$$

$$= -\frac{(H/2)\rho g \cos(\omega t)}{\cosh(kh)} \int_{-h}^{\eta_0} (z+h) \cosh[k(h+z)] dz +$$

$$\rho g \int_{-h}^{\eta_0} z(z+h) \, dz$$

$$= -\frac{(H/2)\rho g \cos(\omega t)}{k^2 \cosh(kh)} \{k(h+\eta_0) \sinh[k(h+\eta_0)] + 1 -$$

$$\cosh[k(h+\eta_0)]\} + \rho g \left(\frac{\eta_0^3}{3} + \frac{h\eta_0^2}{2} - \frac{h^3}{6}\right) \quad (3.12)$$

where the following integral formula is used:

$$\int \theta \cosh(\theta)\, d\theta = \theta \sinh(\theta) - \cosh(\theta) \tag{3.13}$$

The force and moment expressions of Eqs. 3.11 and 3.12, respectively, show that these quantities can be determined by simply measuring the kinematic properties of the wave at the vertical flat barrier.

The *paths* of the water particles beneath the standing wave are determined by using the velocity potential of Eq. 3.9 in Eq. 1.13, so as to introduce the stream function ψ. Thus the Cauchy-Riemann relations become

$$u = \frac{\partial \phi}{\partial x} = \frac{\partial \psi}{\partial z} = \frac{(H/2)gk}{\omega} \frac{\cosh(kh+kz)}{\cosh(kh)} \sin(kx) \sin(\omega t) \tag{3.14}$$

and

$$w = \frac{\partial \phi}{\partial z} = -\frac{\partial \psi}{\partial x} = -\frac{(H/2)gk}{\omega} \frac{\sinh(kh+kz)}{\cosh(kh)} \cos(kx) \sin(\omega t) \tag{3.15}$$

Since the barrier is located at $x=0$, the results from Eq. 3.14 show that $u=0$ at $kx=0, -\pi, -2\pi, \ldots$; whereas $w=0$ where $kx = -\pi/2, -3\pi/2, \ldots$. These results indicate, then, that the water particles in a standing wave remain in cells with quasi boundaries at $x = -\lambda/2, -\lambda, \ldots$.

The paths of the particles correspond to constant values of the stream function, since ψ is tangent to the velocity vector at any point, and, therefore, the locus of points defined by $\psi =$ constant in an Eulerian reference system is the path of the particles. Hence, integrating Eq. 3.14 with respect to z and Eq. 3.15 with respect to x yields the expression for the stream function:

$$\psi = \frac{(H/2)g}{\omega} \frac{\sinh(kh+kz)}{\cosh(kh)} \sin(kx) \sin(\omega t) \tag{3.16}$$

In Figure 3.4, note that a zero value of the stream function defines the flow boundaries, that is, $\psi = 0$ at $x = 0, -\lambda/2, \ldots$ and also at $z = -h$. The streamlines corresponding to the nonzero values of ψ are also shown in the figure.

3.2. WAVES AT A VERTICAL FLAT BARRIER

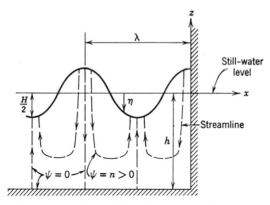

Figure 3.4. Streamlines within a standing wave.

The time-dependent pressure on a concrete sea wall can be destructive. Since the concrete is porous, water is absorbed and lost by the concrete as the free surface of the water rises and falls, respectively. There is a hydrostatic pressure gradient within the wall that results in an alternate push and pull on the concrete particles in the region of wave action. Eventually this pressure cycling causes the wall in this region to crumble and wash away, as illustrated in Figure 3.5.

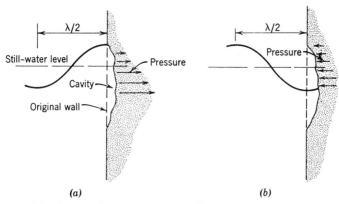

Figure 3.5. Destruction of a porous wall by pressure cycling. (*a*) Crest at wall. (*b*) Trough at wall.

72   FIXED STRUCTURES IN WAVES

Plate 3.1. Areal view of wave refraction at West Hampton Beach on Long Island, New York. (Courtesy of the U.S. Army Coastal Engineering Research Center.)

3.3. WAVES ON A SLOPING FLAT BARRIER

In Chapter 2 it was shown that both the celerity, as described by Eq. 2.24, and the wavelength, in Eq. 2.25, are functions of the water depth. Both of these wave properties decrease with decreasing depth. In this section it is shown that the amplitude or wave height must also vary with depth, increasing as h decreases in value.

Consider an irrotational wave shoaling over a sloping, flat, impermeable barrier, that is, a beach. Following Rayleigh (1877), since the flow is irrotational both the total wave energy and the energy flux are invariant with depth. The energy flux, which is synonymous with *power transmission*, described by Eq. 2.47 is

$$\dot{E} = \frac{\rho a^2 g c_g}{2} = \frac{\rho H^2 g c_g}{8} = \text{constant} \qquad (3.17)$$

where c_g is the group velocity described in Eq. 2.41. Since the group velocity must approach the value of the celerity as the depth of the water decreases, the wave height must increase according to

$$H = 2a = \frac{K}{(c_g)^{1/2}}$$

where K is a proportionality constant. In deep water the wave properties are designated by the subscript zero:

$$K = H_0 (c_{g_0})^{1/2}$$

and, by use of the results of Eq. 2.41, the wave height variation with depth can be described as follows:

$$H = H_0 \left(\frac{c_{g_0}}{c_g}\right)^{1/2} = H_0 \left\{\frac{c_0}{c[1 + 2kh/\sinh(2kh)]}\right\}^{1/2} \qquad (3.18)$$

It must be noted that the wavelength and, therefore, the wave number, k, also vary with depth. To avoid solving a transcendental equation, the ratio of the depth to the wavelength, h/λ, is considered to be an independent variable as presented in Appendix B. The variation of the wave-height ratio, H/H_0, is shown in Figure 3.6 as

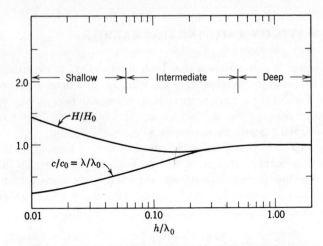

Figure 3.6. Wave property variations in shoaling waters.

a function of h/λ_0. Note the dip in the curve as h/λ_0 decreases. This phenomenon has been experimentally observed by Wiegel (1950) and others.

As $h/\lambda_0 \to 0$, H/H_0 approaches an infinite value. From Section 2.1, in the discussion following Eq. 2.37, however, it is stated that at a depth of $h = a$ in shallow water the wave breaks. Using this breaking condition, one applies the shallow-water approximation to Eqs. 2.24 and 3.18 and combines the results to obtain the following:

$$H_b = H_0 \left(\frac{c_0}{2c_b}\right)^{1/2} = H_0 \left(\frac{c_0}{2\sqrt{gh_b}}\right)^{1/2}$$

where the subscript b refers to the conditions at the break. Rearranging this equation yields for the wave height at the break

$$H_b = 2h_b = H_0^{4/5} \left(\frac{c_0}{\sqrt{2g}}\right)^{2/5} \tag{3.19}$$

Thus the wave height and depth at the break, as predicted by the linear wave theory, depend on the deep-water wave properties.

Now, consider a flat impermeable beach at an angle α to the horizontal, as sketched in Figure 3.7. If the intersection of the still-

3.3. WAVES ON A SLOPING FLAT BARRIER

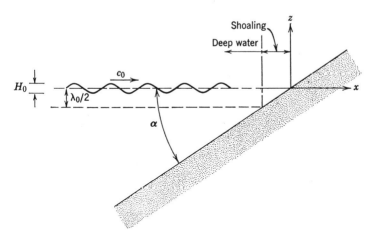

Figure 3.7. Waves on an inclined impermeable flat barrier.

water level and the beach is assumed to be the origin of the coordinate system, the depth of the water at any position is

$$h = x \tan (\alpha) \qquad (3.20)$$

Combining Eqs. 3.19 and 3.20 yields the offshore position of the break:

$$x_b = \frac{H_0^{4/5}}{2 \tan (\alpha)} \left(\frac{c_0}{\sqrt{2g}} \right)^{2/5} \qquad (3.21)$$

When waves approach a beach with crests at an angle to the shore line, the phenomenon of *refraction* occurs. In Figure 3.8 a schematic diagram of an aerial view of shoaling waves is presented. As the celerity of the wave decreases with depth, the following waves begin to "catch up" with the first wave. This causes the wavelength to shorten and the wave front (the crestline) to bend. In the limiting case as $h \to 0$ the wave front aligns itself with the shore line. This phenomenon of wave refraction in water is similar to the wave refraction in both acoustics and optics in that *Snell's law* is obeyed in each case, that is,

$$\frac{\sin (\beta)}{\sin (\beta_0)} = \frac{c}{c_0} \qquad (3.22)$$

76 FIXED STRUCTURES IN WAVES

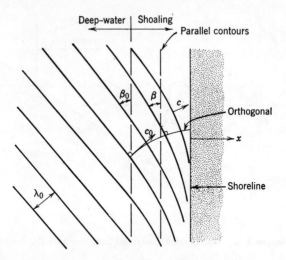

Figure 3.8. Schematic diagram of refracting surface waves.

where (referring to Figure 3.8) β is the angle between the wave front and the shore line. Lines drawn tangent to the celerity or phase velocity are called *orthogonals* since they are normal to the wave fronts. *Contour lines* are lines that pass through points of constant depth on the sea floor. A complete discussion of wave refraction can be found in *Technical Report* No. 4 of the U.S. Army Coastal Engineering Research Center (1966).

3.4. CONSEQUENCES OF VISCOSITY

Although the linear wave theory is adequate for the prediction of the properties of small-amplitude waves, it does not describe the motions of real fluids adjacent to solid boundaries. Because of the property called *viscosity*, all fluids adhere to solid boundaries and therefore offer some resistance to shear stress. The region of viscous action is called the *boundary layer*. Since one of the basic assumptions of the linear wave theory is that the fluid is inviscid, the theory cannot be used to describe the flow within this region. In this section some of the consequences of viscosity that have engineering significance are discussed; however, the analysis of viscous waves is beyond the scope of this book. The text by Kinsman (1965) provides a coverage of the subject.

In Section 3.2 the mechanism of the destruction of a porous sea wall due to hydrostatic pressure cycling is discussed. For a real fluid the wetted surface of the wall also experiences an alternating shear stress caused by a standing wave. Thus the wall particles at the fluid-solid interface are subject to both a vertical shearing force and a pressure force, which contribute to the destruction of the wall.

On the sea floor beneath a standing wave, sand grains are subject to the shear action of the water. The result is a movement of the grains from the region beneath the node of the wave and a deposit of the grains beneath the antinodes. On the sea floor, therefore, a diver observes in the sand symmetric waves similar to those sketched in Figure 3.9. Sand waves or ripples are also caused by steady cur-

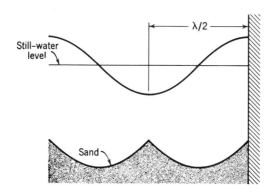

Figure 3.9. Sand waves caused by standing surface waves.

rents passing over the floor. In this case the waves are asymmetric and are sketched as in Figure 3.10.

Another consequence of viscous action is movement of sand and other floor material beneath a shoaling wave. All of these moving materials are referred to as *littoral drift* whereas the actual movement is called *littoral transport*. Two examples of these littoral processes are beach erosion and the creation of a sand bar, that is, an offshore mound of sand submerged in water shallow enough to cause the

78 FIXED STRUCTURES IN WAVES

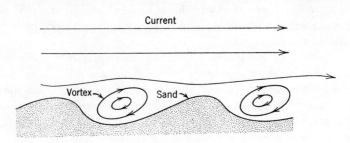

Figure 3.10. Sand waves caused by a steady current.

waves to shoal. Erosion occurs when the net transport of material is away from the beach. A bar is formed when the net transport is into an offshore region.

The extent of the material movement depends on the size and weight of the particles. Thus, in a region of significant wave action or currents, only large rocks may be present since the smaller particles have been washed away. Results of experimental studies show that for a given grain size of sand there is a narrow region of threshold velocities below which no littoral transport occurs. These results and a complete discussion of littoral processes are contained in *Technical Report* No. 4 of the U.S. Army Coastal Engineering Research Center (1966).

3.5. WAVE-INDUCED FORCES ON A PILE

Most offshore structures have as the basic structural element the circular pile. Piles may be vertical, for load bearing, or inclined to withstand severe wind, wave, or current forces. In this section the linear wave theory is used to determine the wave-induced forces and moments on fixed circular piles. The analysis is empirical since the flows in the boundary-layer and wake of the pile, as described in Section 1.5, do not lend themselves to exact mathematical analyses.

Consider the circular pile inclined at an angle α to the horizontal and passing through the free surface in Figure 3.11. The fluid

3.5. WAVE-INDUCED FORCES ON A PILE 79

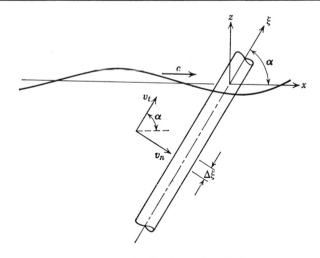

Figure 3.11. An inclined circular pile in waves.

velocity vector at any depth has components both normal to and tangent to the axis of the pile. In the case of Figure 3.11, these velocity components are, respectively,

$$v_n = u \sin(\alpha) - w \cos(\alpha) \qquad (3.23)$$

and

$$v_t = u \cos(\alpha) + w \sin(\alpha) \qquad (3.24)$$

Corresponding to these velocity components are normal forces due to pressure, viscosity, and fluid inertia, and tangential forces that result from the viscous stress on the cylindrical surface. The viscous and pressure forces are usually described in terms of the dimensionless quantity called the *drag coefficient*, which is the ratio of the force in question to the dynamic pressure force of the fluid. As pointed out in Section 1.5 and shown in Figure 1.8, the drag coefficient is found experimentally to be a function of the dimensionless velocity called the *Reynolds number*. Thus, over the elemental length $\Delta \xi$ shown in Figure 3.11, the normal and tangential drag coefficients are, respectively,

$$C_n = \frac{\Delta F_n}{\tfrac{1}{2}\rho v_n^2 D\,\Delta\xi} = f_1\!\left(\frac{v_n D}{\nu}\right) = f_1(\mathrm{R}_D) \qquad (3.25)$$

and

$$C_t = \frac{\Delta F_t}{\tfrac{1}{2}\rho v_t^2 \pi D\,\Delta\xi} = f_2\!\left(\frac{v_t\,\Delta\xi}{\nu}\right) = f_2(\mathrm{R}_\xi) \qquad (3.26)$$

where R_D and R_ξ are the Reynolds numbers based on the pile diameter and elemental length, respectively. Note that the length terms are those that are parallel to the flow directions. The relationships given in Eqs. 3.25 and 3.26 are empirical and are shown in Figure 1.8.

When the angle α is $\pi/2$, $\xi \to z$ and the axial or tangential force is relatively small compared to the normal force. Therefore, in the analysis of the vertical pile, to follow, only the normal force is considered.

The expression for the velocity used in the normal force analysis is derived from the irrotational linear theory and given in Eq. 2.26. The use of Eq. 2.26 appears to be in violation of the concept of irrotationality since viscous losses are now considered. The velocity terms in both Eqs. 3.25 and 3.26, however, are those of the undisturbed flow, which, for all practical purposes, is irrotational. Combining Eqs. 2.26 and 3.25 yields the differential of the normal force on the differential length Δz:

$$\begin{aligned}\Delta F_n &= \tfrac{1}{2}\rho u |u| D C_n\,\Delta z \\ &= \tfrac{1}{2}\rho\left[\frac{ag}{c}\frac{\cosh(kh+kz)}{\cosh(kh)}\right]^2 \cos(kx-\omega t)\,|\cos(kx-\omega t)|\,\cdot \\ &\qquad\qquad DC_n\,\Delta z \qquad (3.27)\end{aligned}$$

where the absolute value of the velocity is needed to preserve the sign variation of the force.

Since the flow is time dependent, the inertial effects of the water must be included in the analysis. The mass of the fluid which is influenced by the cylinder is called the *added-mass* and can be defined in terms of the *added-mass coefficient*:

$$C_i = \frac{\Delta F_i/\Delta z}{\tfrac{1}{4}\rho\pi D^2\,(du//dt)} \qquad (3.28)$$

3.5. WAVE-INDUCED FORCES ON A PILE

where the force ΔF_i is the elemental inertial force of the fluid acting on the pile element. Hydrodynamic theory, such as that contained in the report by Wendel (1956), predicts that the added-mass excited by a circular cylinder in an infinite fluid is equal to the mass of the fluid displaced by the cylinder. Since ΔF_i is equal to the product of the fluid mass and its acceleration, the added-mass coefficient for the circular cylinder is unity. Theoretical values for other cross-sections can be found in the report by Wendel (1956). A selected few of these noncircular cross sections are shown in Table 3.1. According to Ippen (1966), experimental values of C_i for cylinders in waves vary from 0.9 to 2.3 The local acceleration, du/dt, for linear waves is obtained by taking the time derivative of the expression in Eq. 2.26. This result is then combined with Eq. 3.28 to obtain the elemental force on a pile element:

$$\Delta F_i = \tfrac{1}{4}\rho\pi D^2 C_i \left[agk \frac{\cosh\,(kh+kz)}{\cosh\,(kh)} \sin\,(kx - \omega t) \right] \Delta z \quad (3.29)$$

The total normal force on the cylinder is obtained by adding Eqs. 3.27 and 3.29 and integrating the sum over the wetted surface of the cylinder, that is, from $z = -h$ to $z = \eta$. If it is assumed that the center line of the pile is located at $x = 0$ and that the diameter is much smaller than the wavelength, the total wave-induced force on the pile is

$$F = \int_{F(-h)}^{F(\eta)} (dF_n + dF_i)$$

$$= \tfrac{1}{4}\rho \left(\frac{ag}{c}\right)^2 \frac{DC_n}{\cosh^2\,(kh)} \left\{ (h+\eta) + \left(\frac{1}{2k}\right) \sinh\,[2k(h+\eta)] \right\}.$$

$$\cos\,(\omega t) |\cos\,(\omega t)| - \tfrac{1}{4}C_i' \rho\pi D^2 ag \frac{\sinh\,(kh+k\eta)}{\cosh\,(kh)} \sin\,(\omega t) \quad (3.30)$$

where the results of Eq. 2.44 are used in the integration of ΔF_n. Note that the coefficient of added-mass for a circular cylinder is unity, that is, C_i in Eq. 3.30 is 1. The term C_i is shown in that equation for the sake of completeness.

TABLE 3.1

Motion	Cross Section	Added-Mass/ Unit Length	C_i*
	Circle†	$\rho \pi r^2$	1
	Ellipse†	$\rho \pi A^2$	1
	Rectangle‡		
	$A/B = 1$	$1.51 \rho \pi A^2$	1.51
	$A/B = 1/2$	$1.7 \rho \pi A^2$	1.7
	$A/B = 1/5$	$1.98 \rho \pi A^2$	1.98
	$A/B = 1/10$	$2.23 \rho \pi A^2$	2.23
	$A/B = 2$	$1.36 \rho \pi A^2$	1.36
	$A/B = 5$	$1.21 \rho \pi A^2$	1.21
	$A/B = 10$	$1.14 \rho \pi A^2$	1.14
	Flanged Square§		
	$D/A = 0.05$	$1.61 \rho \pi A^2$	1.61
	$D/A = 0.1$	$1.72 \rho \pi A^2$	1.72
	$D/A = 0.25$	$2.19 \rho \pi A^2$	2.19
	Plate†	$\rho \pi A^2$	1

*C_i = added-mass/added-mass for circle.
†Lamb(1945).
‡Lewis(1929).
§Wendel(1956).

3.5. WAVE-INDUCED FORCES ON A PILE

The moment about the base of the pile resulting from the normal force F is the following:

$$M)^+ = -\int_{F(-h)}^{F(\eta)} (h+z)(dF_n + dF_i)$$

$$= -\tfrac{1}{4}\rho\left(\frac{ag}{c}\right)^2 \frac{DC_n}{2\cosh^2(kh)} \left\{\frac{(h+\eta)}{k}\sinh[2k(h+\eta)] + (h+\eta)^2 \right.$$

$$\left. + \left(\frac{1}{2k^2}\right)[1 - \cosh[2k(h+\eta)]]\right\} \cos(\omega t)|\cos(\omega t)| +$$

$$\tfrac{1}{4}C_i\rho\pi D^2 \frac{ag}{\cosh(kh)} \left\{(h+\eta)\sinh[k(h+\eta)]\right.$$

$$\left. - \left(\frac{1}{k}\right)[\cosh[k(h+\eta)] - 1]\right\}\sin(\omega t) \qquad (3.31)$$

The time variations of the free surface, the drag force, and the inertial force are shown in dimensionless form in Figure 3.12. It is seen that F_i leads both F_n and η by $\pi/2$ since the maximum horizontal acceleration occurs when the nodes of the wave pass the structure.

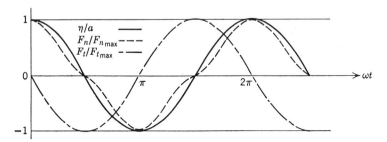

Figure 3.12. Time variation of the forces on a pile and the free-surface displacement for a small amplitude wave.

Now consider an unbraced platform structure comprised of four piles, as sketched in Figure 3.13. The relative magnitudes of the longitudinal separation of the piles and the wavelength determine the magnitude of the net force transmitted to the platform. This fact is illustrated by the following three cases

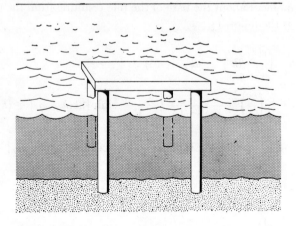

Figure 3.13. Sketch of an unbraced platform in waves.

CASE 1

When the separation of the piles is equal to half the wavelength, a crest will occur at one pile while a trough occurs at the other, as illustrated in Figure 3.14a. Thus the normal drag forces oppose each other, and the net force experienced by the platform is minimum. When two nodes occur at each pile, the drag forces are zero but the inertia forces are maximum and oppose each other; hence, no force is transmitted to the platform. Although the net force in this case is minimum at the platform, the stresses and strains in each pile must both be significant, since the deformation is as shown in Figure 3.14a.

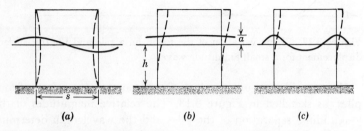

Figure 3.14. Wave-induced deflections of an unbraced platform. (a) $s/\lambda = \frac{1}{2}$. (b) $s \ll \lambda$. (c) $s/\lambda = 1$.

CASE 2

When very long waves are encountered by the structure (i.e., when the separation distance is much smaller than the wavelength), the transmitted force will be maximum since the crest or the node or the trough of this high-energy wave will pass both piles simultaneously, as illustrated in Figure 3.14b, along with the associated deformation of the structure.

CASE 3

When the pile separation and the wavelength are equal, as shown in Figure 3.14c, the transmitted force will be significant, since normal forces induced on each pile will be in phase.

In summary, the wave-induced forces on pile structures are of three types: viscous, pressure, and inertial. For a vertical pile the horizontal forces are the most significant. In this case the viscous and pressure forces lag the inertia of the fluid in time by $\pi/2$. If the pile structure is compound (i.e., consisting of two or more piles). Then the ratio of the pile separation to the wavelength becomes the determining factor in the net force experienced by the structure.

3.6. WAVE-INDUCED VIBRATIONS OF FIXED STRUCTURES

Fixed structures in the ocean experience millions of surface waves each year. Since the waves are of different energies and frequencies, both the *energy spectrum* (i.e., the energy per given frequency) and the *frequency of occurrence* should be determined at the proposed location of a structure. These quantities, which are thoroughly discussed by Kinsman (1965), are illustrated in Figure 3.15. The structure should then be designed so that its natural frequencies are well above the surface wave spectrum in order to avoid a resonance between the wave and structural frequencies.

First, consider a pile of uniform cross section situated in moderately deep water, as illustrated in Figure 3.16. The natural frequencies of the cantilevered structure are given by

$$f_j = \frac{a_i}{2\pi} \left[\frac{EI}{(m+m_w)L^3} \right]^{1/2} \qquad (3.32)$$

86 FIXED STRUCTURES IN WAVES

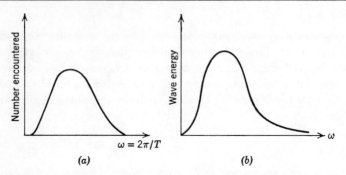

Figure 3.15. Statistical properties of waves. (a) Frequency of wave occurrence. (b) Energy spectrum of waves.

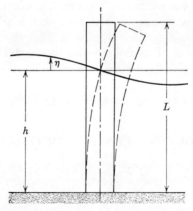

Figure 3.16. Fundamental mode of a pile.

where E is Young's modulus, I is the moment of inertia of the cross-sectional area, L is the length of the pile, m is the mass of the pile, and m_w is the added-mass of the water. Expressions for m_w are presented in Table 3.1 for various cross sections. The modal coefficient a_j, where j is the mode number, has values of 3.52, 22.4, and 61.7 for the first three modes, respectively, according to Crede (1965). Thus the designer of the pile may adjust the length or cross-sectional area or change the material of the pile to ensure that the frequencies obtained from Eq. 3.32 do not exist within the wave spectrum.

3.6. WAVE-INDUCED VIBRATIONS

Now, consider an unbraced platform structure composed of two or more piles, as shown in Figure 3.14. The fundamental frequency of such a structure, according to Crede (1965), is

$$f_1 = \frac{1}{2\pi}\left[\frac{12nEI}{(m+m_w)L^3}\right]^{1/2} \quad (3.33)$$

where m and m_w are the total values and n is the number of piles or legs. Each leg in Figures 3.14b and 3.14c can be considered as two cantilevered beams of length $L/2$. Again, note that the frequency can be raised by reducing the length of each leg. This is accomplished practically by using a brace as shown in Figure 3.17.

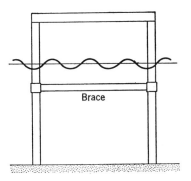

Figure 3.17. A braced platform.

A secondary cause of structural vibrations is the vortex shedding in the wave-induced flow past a pile. Even if the natural frequencies of the structure are above the wave spectrum, these frequencies may still be in the range of those of the vortex shedding. This fact was brought into focus after the destruction of Texas Tower No. 4 off the coast of New England. In the investigation of this disaster [*U.S. Senate Hearings* (1961)], strong evidence was presented which indicated that the failure of the tower was due to resonance between the structural frequency and the shedding frequency of the vortices.

In the case of Figure 3.18, the vortices shed in the wake of a circular pile occur at a frequency f_v, which is a function of both the free-stream velocity, V_n, and the pile diameter, D. The frequency

88 FIXED STRUCTURES IN WAVES

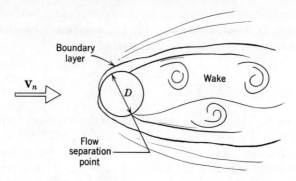

Figure 3.18. Vortex shedding in the wake of a circular pile.

data for this phenomenon are usually presented in dimensionless form by using the *Strouhal number*:

$$S = \frac{f_v D}{V_n} \qquad (3.34)$$

A plot of the Strouhal number as a function of the Reynolds number based on diameter, R_D, is shown in Figure 3.19. This curve approximates the data presented in the paper by Fung (1960).

From the discussion in this section, it should be evident that wave-induced vibrations must be considered in the design of offshore structures. As stated above, even if the frequencies of the surface

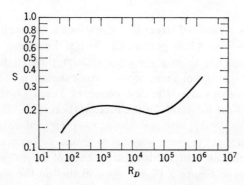

Figure 3.19. Strouhal number for vortex shedding in the wake of a circular cylinder.

waves are below the natural frequencies of the structure, the vortex shedding frequencies may be in the range of the structural frequencies, resulting in an unwanted resonance condition.

3.7. WAVE-MAKING DRAG

A problem encountered by both ocean engineers working in swift estuarine waters and naval architects is *wave-making drag*, that is, the drag resulting from the creation of surface waves by an object in a stream or by a vehicle traveling in still water. To obtain an idea of the wave patterns caused by a moving disturbance, consider a pile in a deep-water stream as illustrated in Figure 3.20. The presence of

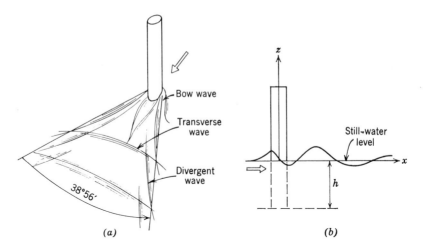

Figure 3.20. Waves in the wake of a pile. (*a*) Waves created by a pile in a stream. (*b*) Free-surface profile for the transverse waves.

the pile causes both local disturbances in the forms of a *bow wave* and *divergent waves* and a group of *transverse waves* that are called *free waves*. All of these waves remain with the pile at relative positions that depend on the velocity of the stream. With reference to Figure 3.20, the amplitudes of the transverse waves decrease in the downstream direction since the constant energy of each wave is distributed

90 FIXED STRUCTURES IN WAVES

over longer crest lengths. The lines drawn through the intersections of the transverse and divergent waves are theoretically separated by an angle of 38°56′ [Lamb (1945)]. The wave pattern shown in Figure 3.20 is called the *Kelvin wave pattern* because of the original work of Lord Kelvin (Sir W. Thomson) on the subject in 1887.

To analyze both the wave properties and the resulting drag force on the pile, assume that the pile is moving at a constant velocity V in still water, as shown in Figure 3.21. As previously mentioned, the transverse and divergent waves have fixed positions with respect to

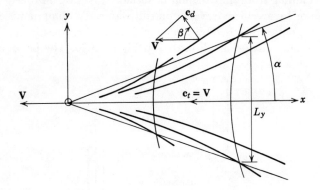

Figure 3.21. The Kelvin wave pattern.

the pile, and, therefore, their celerities have components in the direction of the pile motion equal to the pile velocity. If the water is assumed to be deep, so that $\tanh(kh) \to 1$ in Eq. 2.24, the celerities of the transverse and divergent waves are, respectively,

$$c_t \simeq V = \left(\frac{g}{k_t}\right)^{1/2} = \left(\frac{g\lambda_t}{2\pi}\right)^{1/2} \tag{3.35}$$

and

$$c_d = V\cos\beta = \left(\frac{g}{k_d}\right)^{1/2} = \left(\frac{g\lambda_d}{2\pi}\right)^{1/2} \tag{3.36}$$

Thus the pile velocity determines the wavelengths of the various patterns, so that the distance between the crests increases as the velocity of the pile increases.

3.7. WAVE-MAKING DRAG

In shallow water Eq. 2.24 is approximately $c = (gh)^{1/2}$, and the resulting wave pattern is in the Kelvin form only for pile velocities less than 40% of the celerity, that is, $V/c < 0.40$ [Havelock (1908)]. Above this value the angle α defining the wave envelope increases and approaches the value $\pi/2$ as the pile velocity approaches the celerity. When $V/c = 1$, the condition is called *critical* and is analogous to the situation of a body traveling at the speed of sound in air. When $V/c > 1$, the transverse waves vanish, leaving only the divergent waves, and the angle of the wave envelope again decreases according to the Mach-type relation

$$\frac{V}{c} = \frac{V}{(gh)^{1/2}} = \frac{1}{\sin(\alpha)} \qquad \frac{V}{c} > 1 \qquad (3.37)$$

The variation of α with the Froude number based on depth is shown in Figure 3.22. It is noted here that a disagreement exists among wave resistance experts as to the shape of the divergent waves in the region defined in Eq. 3.37; some believe that the waves are concave, whereas others think that they are convex. This disagreement, which is noted by Wigley in the collection of Havelock's papers (1965), is not germane to this discussion since the large majority of problems encountered by ocean engineers are in the subcritical flow region.

The theoretical analysis of wave-making resistance is beyond the scope of this book because of its complexity. To acquaint the reader with the basic concepts of the analysis, however, the fundamentals as presented by Havelock in his 1934 paper are reviewed.

The creation of a wave pattern by a body such as a ship or a pile is due to a transfer of energy from the body to the fluid if the body is in motion. When a fixed body, such as a pile, is placed in a stream, the momentum of the stream is redirected because of the presence of the pile and a new energy distribution takes place within the liquid in the form of surface waves.

Consider the transverse waves in deep water shown in Figure 3.21. These waves would normally travel in a group with a group velocity c_g; as described by Eq. 2.41. The moving body, however, forces the group to travel at the higher velocity V. Hence the power required to maintain the higher group velocity must be the difference between the power of the wave system at the velocity V, as represented by the energy flux described in Eq. 2.47, and that at the natural group

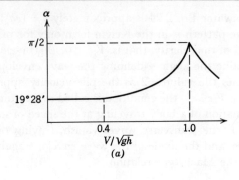

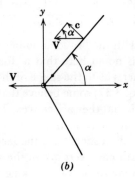

Figure 3.22. Shallow-water wake. (a) Angular variation with Froude number based on depth. (b) Supercritical wave envelope.

velocity c_g. Mathematically, if $\mathbf{F}_R$ is the drag force acting on the body and $\mathbf{L}_y$ is a horizontal line across which a wave passes normally, then the power lost is

$$\mathbf{F}_{\dot{R}} \cdot \mathbf{V} = [\dot{\mathbf{E}}(V) - \dot{\mathbf{E}}(c_g)] \cdot \mathbf{iL}_y \qquad (3.38)$$

From the discussion of the wave group in Section 2.2, it is evident that the wave drag in shallow water, where the celerity and group velocity are equal, must be zero, following the reasoning leading to Eq. 3.38.

Now, consider both types of waves in the Kelvin wave pattern. According to Havelock (1934), this pattern is created by the superposition of an infinite number of regular waves of equal amplitude but with celerities that depend on their direction according to

3.7. WAVE-MAKING DRAG

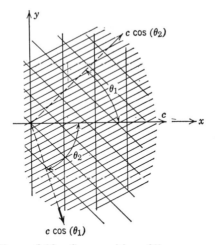

Figure 3.23. Superposition of linear waves.

$[c \cos (\theta)]$, referring to Figure 3.23. The displacement of the free surface at any point (x, y) is then

$$\eta(x,y) = \int_{-\pi/2}^{\pi/2} A(\theta) \sin [(kx \cos \theta + ky \sin \theta) \sec^2 \theta] d\theta \quad (3.39)$$

where the amplitude term, $A(\theta)$, depends on the shape of the body creating the waves. It should be noted here that the time term for traveling waves does not appear in Eq. 3.39 since the waves are always at fixed positions relative to the body. Using the expression in Eq. 3.38, Havelock (1934) determined the wave-making drag on the body to be

$$F_R = \pi \rho V^2 \int_0^{\pi/2} [A(\theta)]^2 \cos^3 \theta \, d\theta \quad (3.40)$$

For the simplest of body shapes the evaluation of the integrals in Eqs. 3.39 and 3.40 is most difficult; however, to illustrate the results obtained from the theory let $A(\theta)$ be a constant, C, which is a function only of the velocity, V. This amplitude function could be considered to that of a point disturbance, that is, a body having no length or breadth. For a body of finite length and breadth, the integrals in Eqs. 3.39 and 3.40 would contain a number of terms that would be functions of the angle θ and would represent the contributions to the wave pattern and drag from the bow, the middle body points, and the stern of the body. Continuing with the example, Havelock states in his 1934 paper that the contribution to the drag from the transverse wave is primarily in the region from $0 < \theta < 35°16'$, while that of the divergent waves comes from the region $35°16' < \theta < \pi/2$. Hence the wave-making resistance on this hypothetical body is

$$\begin{aligned} F_R &= F_R(\text{transverse}) + F_R(\text{divergent}) \\ &= \pi \rho V^2 C^2 \left(\int_0^{35°16'} \cos^3 \theta \, d\theta + \int_{35°16'}^{\pi/2} \cos^3 \theta \, d\theta \right) \\ &= \pi \rho V^2 C^2 (0.5131 + 0.1535) \\ &= \pi \rho V^2 C^2 (\tfrac{2}{3}) \end{aligned} \qquad (3.41)$$

From the results in Eq. 3.41 it is evident that the transverse waves are responsible for approximately 77% of the wave drag, while the divergent waves contribute only 23%.

Although the disturbance used in the illustration is hypothetical, the relative contributions to the wave resistance from each wave system are in the neighborhoods of those obtained for most body shapes found in engineering situations. As previously mentioned, the various parts of a finite body contribute to the energy in the trailing waves and, therefore, to the wave-making resistance. Although these contributions do affect the energy distribution, as represented by the wave amplitude, they do not alter the wave pattern itself.

Experimental and theoretical data on wave resistance are presented in the form of a drag coefficient as a function of the Froude number based on body length L, that is, C_R versus $V/(gL)^{1/2}$. An example of this for a ship is shown in Figure 3.24. The humps in the curve exist because the wave drag on the body depends on the relative body positions of the crests and troughs of the bow and stern waves. A discussion of this phenomenon and more detailed analyses of wave-making resistance can be found in the collection of Sir Thomas Havelock's papers edited by Wigley (1965), the classical paper of Inui (1962), and the doctoral thesis of Ward (1962).

3.8. EXPERIMENTAL STUDIES

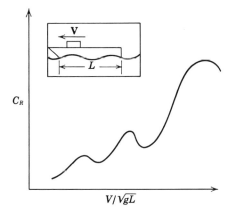

Figure 3.24. Typical ship wave resistance.

3.8. EXPERIMENTAL STUDIES

Equipped with a basic knowledge of the theories of wave-induced forces and motions of fixed structures, the reader is now introduced to two experimental studies of the phenomena. This introduction, coupled with material presented in the preceding sections, should provide a good understanding of wave action upon structures.

a. Wave-Induced Forces and Motions

A pine model of an unbraced off-shore platform, such as that sketched in Figure 3.13, is placed in a wave tank. The model, shown schematically in Figure 3.25, has a fundamental frequency of 2.82 Hz when dry. After being placed in the water the fundamental frequency is reduced to 2.15 Hz because of the added-mass of the water. By knowing the dimensions of the model and Young's modulus (1600×10^3 psi), the added-mass is found to be 1.18 slugs from Eq. 3.33. Since the effective frequency range of the wave-maker in the tank is 0.9 to 2.0 Hz, a weight of 100 lb is placed on the platform to lower the fundamental frequency to a value of 1.47 Hz.

With the platform free to vibrate, waves of various frequencies but of constant amplitude are generated. The resulting longitudinal

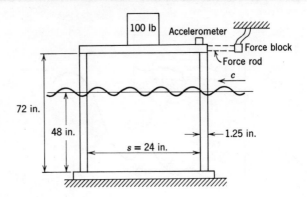

Figure 3.25. Wave-induced force and motion apparatus.

and transverse motions of the structure are then measured by using two accelerometers with sensitivities of 5 g/V. The signals from the accelerometers are recorded on a six-channel strip-chart recorder, along with signals from two resistance wire wave gauges. One of the wave gauges is located at the leading leg or legs, depending on the orientation of the structure, and the other gauge is placed 6 ft downstream to minimize interference.

The first set of data is obtained with one side of the square platform aligned with the wave front, as shown in Figure 3.26a. The structure is then rotated through an angle of 30° for the second set of data, as shown in Figure 3.26b.

Finally, with the structure as shown in Figure 3.26a, the platform is rigidly connected to a force block on a towing carriage to measure the wave-induced forces transmitted to the platform.

The results of the vibration and force tests are shown in Figure 3.27 in the form of the acceleration response as a function of the wave frequency. When the platform and wave front are aligned, the longitudinal motions are much greater than the transverse motions, as expected. Resonance occurs at the fundamental frequency of the structure for both modes. The finite amplitudes at the resonant frequency, as well as the steepness of the response curves about this frequency, are determined by the damping in the system. A detailed analysis of the effects of damping is given by Crandall and Mark (1963).

3.8. EXPERIMENTAL STUDIES

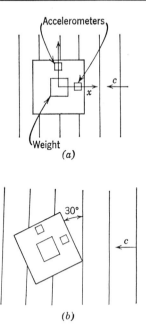

Figure 3.26. Platform orientation. (*a*) Platform and wave front aligned. (*b*) Platform rotated 30°.

When the structure is rotated 30°, the acceleration response of the transverse mode increases while that of the longitudinal mode decreases. In addition, there is a torsional motion, which is detected by stretching a horizontal wire just above the platform and noting the angle between the wire and an initially parallel line drawn on the platform. The frequency of the torsional vibration is approximately 2 Hz.

In regard to the linear modes, for a given frequency an alternating increase and decrease in the acceleration response of each mode is noted. This phenomenon, which is called *beating*, occurs when there is an alternating transfer of energy from one vibrational mode to another, that is, between transverse and longitudinal or between linear and torsional. Beating occurs in this situation because the center of mass of the structure is not coincident with the geometric center line.

98 FIXED STRUCTURES IN WAVES

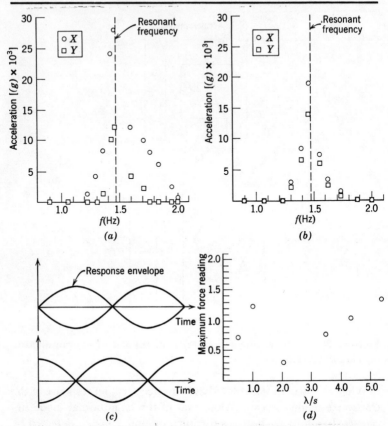

Figure 3.27. Results of the wave force and motion experiment. (a) Platform and wave front aligned: $H = 1.0$ in. (b) Platform and wave front at $30°$: $H = 1.0$ in. (c) Beating. (d) Wave-induced force: $H \simeq 1.0$ in.

From the results of this experiment the reader can see that it is not adequate to design a structure from strength considerations alone. In addition, the vibrational properties and the mass distribution must be taken into account.

b. Wave-Making Resistance

The phenomenon of wave-making resistance or drag is studied by towing three circular cylinders at various speeds in still water. The

3.8. EXPERIMENTAL STUDIES

cylinders are supported above the water on the towing carriage by connecting them to a force block. Small end plates must be attached to the submerged end of the shaft being towed to prevent axial flows near that end. These axial flows result in undesired nonuniform tip losses. A sketch of the resistance apparatus is shown in Figure 3.28.

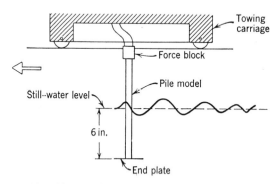

Figure 3.28. Sketch of wave resistance experimental apparatus.

The three pine shafts used in the study have respective diameters of $\frac{3}{8}$, $\frac{5}{8}$, and $\frac{13}{16}$ in, but all are of the same length and draught. As previously mentioned, attached to the submerged end of each cylinder is a 2-in.-square thin metal plate, as shown in Figure 3.28. The towing speeds ranged from 1.1 to 4.4 fps.

The force blocks are used to measure the total resistance on each cylinder, which is the sum of the viscous-pressure drag, discussed in Section 1.5, and the wave-making drag. The results of the experiment are presented in dimensional form in Figure 3.29 and in dimensionless form in Figure 3.30. The dimensionless variables shown in Figure 3.30 are the drag coefficient and the Froude number based on diameter, that is, the functional relationship shown is

$$C_j = \frac{F_j}{\tfrac{1}{2}\rho V^2 DL} = f\left[\frac{V}{(gD)^{1/2}}\right] \tag{3.42}$$

Shown in Figure 3.30 are the total drag coefficient, C_T; the normal (viscous-pressure) drag coefficient, C_n, which is obtained from Figure 1.8; and the wave-drag coefficient C_R, which is equal to the difference $C_T - C_n$. The data in dimensionless form are somewhat misleading

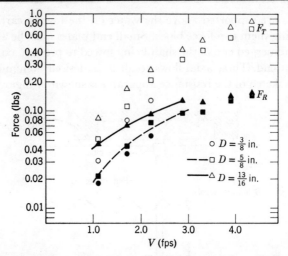

Figure 3.29. Total force and wave drag on the cylinders.

since both C_T and C_n decrease with increasing Froude number, whereas the corresponding drag forces both increase markedly with towing speed. The decreasing drag coefficients are due to the V^2 term in the denominator of Eq. 3.42.

At speeds above 3.0 fps the shaft of smallest diameter begins to vibrate because of the vortex shedding in its wake. For this reason only three of the resistance data are shown in the figures for the $\frac{3}{8}$-in. diameter cylinder.

It is interesting to study the results obtained from Eq. 3.41 when that theoretical expression is applied to the experiment. Fitting curves to the force-speed data of Figure 3.29 yields the following empirical relationships:

$$F_R = 0.0182 V^{1.56}; \quad \tfrac{5}{8}\text{-in. diameter}$$

and

$$F_R = 0.0407 V^{1.07}; \quad \tfrac{13}{16}\text{-in. diameter}$$

When these expressions are compared with Eq. 3.41, the following are obtained:

$$C = 0.0655 V^{-0.220}; \quad \tfrac{5}{8}\text{-in. diameter}$$

and

$$C = 0.0970 V^{-0.465}; \quad \tfrac{13}{16}\text{-in. diameter}$$

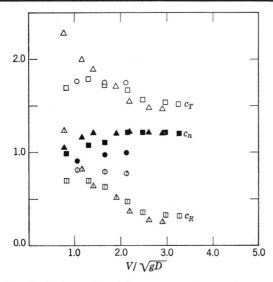

Figure 3.30. Coefficients of total drag, normal drag, and wave drag on the cylinders.

Thus the parameter C for this speed range is a relatively weak function of the towing speed. Curves obtained from these empirical expressions are shown in Figures 3.29.

Finally, although the viscous-pressure resistance and the wave-making resistance are both given in Figure 3.30, it must be remembered, when applying experimental results like those presented herein to a prototype, that Reynolds scaling and Froude scaling cannot be performed simultaneously.

3.9. ILLUSTRATIVE EXAMPLES

a. Wave-Induced Forces and Moments

Consider a 1-ft-diameter wooden pile which is 20 ft in length and is situated in 10 ft of water. The weight density of the pile is 30 lb/ft^3, and Young's modulus is 16×10^5 psi. The pile is subject to 1-ft waves the length of which is 200 ft. The maximum wave-induced

102 FIXED STRUCTURES IN WAVES

force on the pile and the corresponding moment about the base are to be determined, along with the possible resonance condition.

First, the properties of the waves must be determined. Since $h/\lambda = \frac{1}{20}$ is in the shallow-water range, so that the hyperbolic functions are approximately $\sinh(kh) \simeq \tanh(kh) \simeq kh$ and $\cosh(kh) \simeq 1$, Eq. 2.24 becomes

$$c = (gh)^{1/2} = 17.94 \text{ fps}$$

Thus the wave frequency is $f = c/\lambda = 0.08975$ Hz.

The force and the moment on the pile resulting from the wave are obtained from the expressions in Eqs. 3.30 and 3.31. These equations applied to the shallow-water structure are, respectively,

$$F = \tfrac{1}{4}\rho\left(\frac{ag}{c}\right)^2 DC_n 2(h+\eta)\cos(\omega t)|\cos(\omega t)|$$
$$- \tfrac{1}{4}\rho agk\pi D^2 C_i (h+\eta)\sin(\omega t)$$

and

$$M = -\tfrac{1}{4}\rho\left(\frac{ag}{c}\right)^2 DC_n \tfrac{3}{2}(h+\eta)^2\cos(\omega t)|\cos(\omega t)| +$$
$$\tfrac{1}{4}\rho agk\pi D^2 C_i (h+\eta)^2 \sin(\omega t)$$

If the depth of the water is larger than the amplitude of the wave by at least an order of magnitude, then the time-dependent free-surface displacement $\eta(t)$ can be neglected in the equations with acceptable error. In the present example $a/h = \frac{1}{20}$, so that the error involved in the force computation is about 5%.

Now, if it is assumed that $\rho = 2.0$ slugs/ft³, $C_n = C_i = 1$, and η is negligible, the force expression becomes

$$F = 8.05 \cos(\omega t)|\cos(\omega t)| - 7.95 \sin(\omega t)$$

To find the maximum force the inflection points in the force-time curve must be determined. Thus, setting the time derivative of the equation equal to zero, one finds that the point of maximum force occurs when $\omega t = 0.8360\pi$ rad or 1.836π rad; therefore $|F_{max}| = 10.01$ lb and $|M_{max}| = 53.7$ lb-ft. Referring to Figure 3.12, one sees that the maximum force (and moment) on the pile occurs just after the passing of a wave node.

It is important to note that the viscous-pressure force is a strong function of the amplitude, being proportional to a^2, whereas the inertial force is proportional to a.

The fundamental frequency of the pile is obtained from Eq. 3.32, that is,

$$f_1 = \frac{a_1}{2\pi}\left[\frac{EI}{(m+m_w)L^3}\right]^{1/2}$$

where $m = 14.60$ slugs and $m_w = 15.70$ slugs. Hence, with $I = \pi D^4/64$ and $a_1 = 3.52$, the fundamental frequency is 3.40 Hz. Comparing this structural frequency to that of the wave, one can conclude that resonance will not occur.

Now, the vortex-shedding frequency in the wake of the pile must be determined. As shown in Section 3.6, the dimensionless frequency, called the *Strouhal number*, is a function of the Reynolds number. The maximum horizontal particles velocity experienced by the pile is found by applying Eq. 2.26 to the shallow-water situation:

$$u_{max} = \frac{ag}{c} = 0.897 \text{ fps}$$

Hence the maximum Reynolds number is

$$R_D = \frac{u_{max}D}{\nu} = 8.57 \times 10^4$$

The corresponding Strouhal number is now found in Figure 3.19:

$$S = \frac{f_v D}{u_{max}} = 0.2$$

The vortex-shedding frequency is then $f_v = 0.179$ Hz, a value well below that of the fundamental frequency of the structure; therefore the pile will not be excited by either the waves or the vortices being shed in its wake.

b. Shoaling Waves and Wave Refraction: Computer Solution

A structure is to be located approximately 100 ft offshore. The beach, which has a slope of 0.10, is on the east coast and is subject to waves coming in directly out of the northeast. With reference to

Figure 3.8, the length, celerity, height, and direction of the waves at the structure must be determined. The deep-water waves of interest are those having lengths and heights in the ranges 100 ft $\leq \lambda_0 \leq$ 1000 ft and 1 ft $\leq H_0 \leq$ 10 ft, respectively. Because of both the iterations involved in the wavelength calculations and the large number of height-wavelength combinations involved, a digital computer is used and the applicable equations are programmed in FORTRAN IV.

Since this situation involves wave refraction as illustrated in Figure 3.8, the angular variation of the wave must be determined. The two-dimensional theory can be considered to be that describing the projection of a three-dimensional wave on the $x - z$ plane; for example, the group velocities in Eq. 3.18 can be replaced by

$$c_{gx} = c_g \cos(\beta) \tag{3.43}$$

and

$$c_{gx_0} = c_{g_0} \cos(\beta_0) \tag{3.44}$$

where the subscript x refers to the two-dimensional group velocity in the $x - z$ plane, and the subscript 0 refers to that in deep water. Hence, to include the wave refraction, one simply replaces the group velocities in Eq. 3.18 by the expressions in Eqs. 3.43 and 3.44 to obtain the wave height at any offshore position, that is,

$$\begin{aligned} H &= H_0 \left(\frac{c_{gx_0}}{c_{gx}}\right)^{1/2} \\ &= H_0 \left(\frac{c_{g_0}}{c_g}\right)^{1/2} \left[\frac{\cos(\beta_0)}{\cos(\beta)}\right]^{1/2} \end{aligned} \tag{3.45}$$

where the term $[\cos(\beta_0)/\cos(\beta)]^{1/2}$ is called the *refraction coefficient*. Since neither c_g (and, therefore, c) or the angle β is known, a second equation relating these quantities is needed. That equation is the expression of Snell's law:

$$\frac{\sin(\beta)}{\sin(\beta_0)} = \frac{c}{c_0} = \frac{\lambda}{\lambda_0} \tag{3.46}$$

Physically, Snell's law states that the separation of the wave crests in the y direction (the direction parallel to the shore line) is independent of the depth of the water.

3.9. ILLUSTRATIVE EXAMPLES

The variation of the wavelength with depth is obtained from Eq. 2.25, that is,

$$\lambda = \left(\frac{gT^2}{2\pi}\right) \tanh\left(\frac{2\pi h}{\lambda}\right) \qquad (2.25)$$

which must be solved numerically as described in Example 2.6c.

Equations 3.45 and 2.41 can now be combined along with Eq. 3.46 to obtain the following wave-height expression:

$$H = H_0 \left\{\frac{c_0}{c[1 + 2kh/\sinh(2kh)]}\right\}^{1/2} \left[\frac{\cos(\beta_0)}{\sqrt{1 - (c/c_0)^2 \sin^2(\beta_0)}}\right]^{1/2} \qquad (3.47)$$

The procedure followed in the computer solution of the problem is shown in the flow diagram of Figure 3a. The FORTRAN IV program is the following:

```
      READ 100, PI, G, H
      READ 150, HTO, BETA, XO
   10 CO = SQRT(G*XO/(2.*PI))
      T = XO/CO
      PRINT 200, XO, T
      X = 1.
   20 Y = 2.*PI*H/X
      TANHY = (EXP(Y) − EXP(−Y))/(EXP(Y) + EXP(−Y))
      Z = G*T**2.*TANHY/(2.*PI)
      IF(ABS(Z − X) − 0.001)40, 40, 30
   30 X = Z
      GØ TØ 20
   40 PRINT 250, Z
      C = X/T
      PRINT 250, C
      P = SQRT(1.−(C/CO)**2.*SIN(BETA))
      SINH2Y = (EXP(2.*Y) − EXP(−2.*Y))/2.
      Q = (1 + 2.*Y/SINH2Y)
   50 HT = HTO*(CO*COS(BETA)/(C*P*Q))**0.5
      PRINT 250, HT
      IF(HTO − 10.)60, 70, 70
   60 HTO = HTO + 1.
      GØ TØ 50
```

```
70    XO = XO + 10.
      IF(XO − 1000.)80, 90, 90
80    GØ TØ 10
90    STØP
100   FØRMAT(3E12.6)
150   FØRMAT(3E12.7)
200   FØRMAT(2F12.7)
250   FØRMAT(F12.7)
```

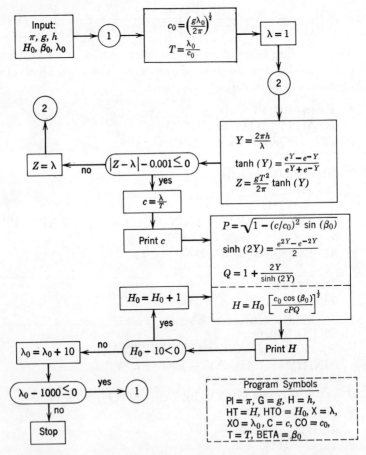

Figure 3a. Flow diagram of computer analysis of refraction.

3.9. ILLUSTRATIVE EXAMPLES

To check the program, a sample deep-water wave having a height of 1 ft and a length of 100 ft is studied. The wavelength at the structure is found to be 71.0 ft, a result obtained by using a graphical solution of Eq. 2.25. Since the period does not vary with the depth, the celerity is found to be $c = 16.05$ fps, and the angle between the crest and the bottom contour is $\beta = \sin^{-1}(0.50) = 30°$. Finally, Eq. 3.47 yields a wave height of 0.845 ft.

It is interesting to note that, if the wave approached the shoreline directly out of the east ($\beta_0 = 0°$), the wave height at the structure, from Figure 3.6, would be 0.933 ft, that is, in the range of intermediate depth, where the height of the wave is less than its deep-water height. Comparing these two heights shows that the refraction process alters the shoaling behavior of the wave properties.

c. Wave-Making Resistance

A bridge is to be erected across the mouth of a river having a maximum current of 10 fps and a depth of 10 ft. Piles of 1-ft diameter are to be used in the structure, each having a length of 20 ft above the river bed. From the experimental results presented in Figure 3.30, the maximum force and the bending moment about the base on each pile are to be determined. It is assumed that the separation of the piles is sufficient to prevent hydrodynamic interference between them.

The *Froude number* based on pile diameter for the maximum current velocity is

$$\frac{V}{(gD)^{1/2}} = 1.765$$

From Figure 3.30, the coefficient is approximately 0.58, that is,

$$\frac{F_R}{\frac{1}{2}\rho V^2 D L_{\text{wet}}} = 0.58$$

so that the wave drag is $F_R = 580$ lb. This force acts approximately at the position on the pile corresponding to the position of the average horizontal particle velocity in the transverse waves in the wake of the pile (see Figure 3.20).

108　FIXED STRUCTURES IN WAVES

Before determining the position of the wave drag, the length of the transverse waves must first be calculated from Eq. 3.35. Thus

$$\lambda_t = \frac{2\pi V^2}{g} = 19.5 \text{ ft}$$

If it is assumed that the wave influences the water particles to a depth of $\lambda/2$, the assumption of deep water made in the derivation of Eq. 3.35 appears to be valid, since the depth is greater than half the wavelength. Applying the deep-water approximation to Eq. 2.26 yields, for the average horizontal velocity of the particles within the wave,

$$\bar{u} = \frac{2}{\lambda} \int_{-\lambda/2}^{0} u \, dz = \frac{agk}{\omega} \frac{2}{\lambda} \int_{-\lambda/2}^{0} e^{kz} \, dz \cos(kx - \omega t)$$

$$= \frac{2ag}{\omega\lambda}(1 - e^{-\pi}) \cos(kx - \omega t) \tag{3.48}$$

This velocity is found at a depth of

$$\bar{z} = \left(\frac{1}{k}\right) \ln\left[\frac{2(1 - e^{-\pi})}{\lambda k}\right] \tag{3.49}$$

$$= -3.69 \text{ ft}$$

The moment of the wave drag about the base of the pile is

$$M_R)^+ = -F_R(h + \bar{z})$$
$$= -3,660 \text{ lb-ft}$$

The viscous-pressure force must be determined using Figure 1.8. The C_n values in Figure 3.30 cannot be used, since Froude scaling and Reynolds scaling cannot be done simultaneously. The Reynolds number based on the pile diameter, assuming a kinematic viscosity of 1.06×10^{-5} ft²/sec, is $R_D = 9.44 \times 10^5$. The corresponding value of C_n is then 0.32, which yields a viscous-pressure drag of $F_n = 320$ lb. This force is assumed to act at a depth of half the wetted length of the pile; therefore the moment about the base is 1600 lb-ft.

Finally, the total moment about the base is the sum of the two components, that is, $M_T = M_R + M_n = -5,260$ lb-ft.

3.10. REFERENCES

Bagnold, R. A. (1939), "Interim Report on Wave-Pressure Research," *J. Inst. Civil Engrs.*, June, pp. 202–226.

Crandel, S., and Mark, W. (1963), *Random Vibrations in Mechanical Systems*, Academic Press, New York.

Crede, C. E. (1965), *Shock and Vibration Concepts in Engineering Design*, Prentice-Hall, Englewood Cliffs, N.J.

Fung, Y. C. (1960), "Fluctuating Lift and Drag Acting on a Cylinder in a Flow at Supercritical Reynolds Numbers," *Inst. Aeronaut. Sci.*, Paper No. 60-6.

Havelock, T. H. (1908), "The Propagation of Groups of Waves in Dispersive Media, with Application to Waves on Water Produced by a Travelling Disturbance," *Proc. Roy. Soc.*, A, **81**, 398–430.

——— (1934), "Wave Patterns and Wave Resistance," Institute of Naval Architects, Summer Meeting, July 12.

Inui, T. (1962), "Wave-Making Resistance of Ships," Society of Naval Architects and Marine Engineers, Annual Meeting, November 15–16.

Ippen, A. T., Ed. (1966), *Estuary and Coastline Hydrodynamics*, McGraw-Hill, New York.

Kinsman, B. (1965), *Wind Waves*, Prentice-Hall, Englewood Cliffs, N.J.

Lamb, H. (1945), *Hydrodynamics*, Dover Publications, New York.

Lewis, F. (1929), "The Inertia of Water Surrounding a Vibrating Ship," *Trans. Soc. Naval Architects Marine Engrs.*, **37**.

Rayleigh, (1877), "On Progressive Waves," *Proc. London Math. Soc.*, **9**, 21–26.

Thomson, W. (Lord Kelvin) (1887), "On Ship Waves," *Proc. Inst. Mech. Engrs.*, *Popular Lectures and Addresses*, iii. 450.

U.S. Army Coastal Engineering Research Center (1966), "Shore Protection, Planning and Design," *Tech. Rept. No. 4*, 3rd ed., U.S. Government Printing Office, Washington, D.C.

U.S. Senate Hearings (1961), p. 195.

Ward, L. W. (1962), "A Method for the Direct Experimental Determination of Ship Wave Resistance," Doctoral Dissertation, Stevens Institute of Technology, Hoboken, N.J.

Wendel, K. (1956), "Hydrodynamic Masses and Hydrodynamic Moments of Inertia," *David Taylor Model Basin Transl. No. 260*, July.

110 FIXED STRUCTURES IN WAVES

Wiegel, R. (1950) "Experimental Study of Surface Waves in Shoaling Water," *Trans. Am. Geophys. Union*, pp. 377–385.

Wigley, C., ed. (1965), *The Collected Papers of Sir Thomas Havelock on Hydrodynamics*, U.S. Government Printing Office, Washington, D.C.

3.11. PROBLEMS

1. A standing wave 10 ft long and 3 ft high exists at a sea wall in 6 ft of salt water. Plot the pressure distribution on the wall when
   (a) a crest is at the wall;
   (b) a trough is at the wall.

2. Determine the time-dependent force and moment expressions at the sea wall described in Problem 1.

3. If a 20-ft-long standing wave 6 in. in height exists in 5 ft of salt water, what are the maximum values of
   (a) the vertical velocity component of the fluid particles adjacent to the sea wall;
   (b) the horizontal velocity component of the fluid particles adjacent to the floor?

4. Sketch the stream functions corresponding to $\psi = 0$, 1, and 2 for the wave described in Problem 1.

5. A deep-water wave 5 ft in height approaches the east coast of an island from the east. A digital wave staff located in 10 ft of water measures the height of the shoaling wave. At this position $H = 4.6$ ft. Determine the length and celerity of the wave both as it passes the staff and when it is in deep water.

6. If the beach angle in Problem 5 is 0.10 rad. how far offshore will the wave break?

7. Assume that the wave in Problems 5 and 6 approaches directly out of the northeast. What angle will the wave make with the bottom contours at the wave staff?

8. Write the expressions for the normal and tangential velocity components on a deep-water pile that is both inclined at an angle α and subject to linear waves.

9. A 9-in.-diameter vertical pile is located in 30 ft of water and is subject to 2-ft waves of 20-ft length. Determine the maximum

force on the pile and the maximum moment about the base, assuming $\rho = 2.00$ slugs/ft^3 and $\nu = 1.05 \times 10^{-5}$ ft^2/sec.

10. For a vertical pile located in shallow water determine the expressions for the following:
   (a) the *loading function*, that is, $w(z, t) = -\partial F_s/\partial z$;
   (b) the *shear force*, that is, $F_s(z, t) = \partial M_b/\partial z$;
   (c) the *bending moment*, that is, $M_b(z, t)$;
   (d) the *deflection of the pile's center line*, that is, $X = X(z, t)$, assuming that $\partial^2 X/\partial z^2 = M_b/EI$.

11. A 9-in.-diameter pile is located in 15 ft of water and is subject to 3-ft waves that are 20 ft in length. The material of the pile has a weight density of 0.03 lb/in^3., and the pile itself is 20 ft in total length. If $E = 1.6 \times 10^6$ psi and $I = \pi D^4/64$, will resonance with the wave or vortex shedding occur?

12. Four piles like the one described in Problem 11 support a square platform and equipment weighing 2 tons.
   (a) What is the fundamental frequency of the structure?
   (b) Where must light-weight braces be applied to double this frequency?

13. What is the maximum vortex shedding frequency for the waves experienced by the pile in Problem 11 if the diameter is
   (a) doubled;
   (b) halved?

14. A periscope extends above a submerged submarine through the free surface. The submarine is traveling at a speed of 10 knots. Assuming deep-water conditions, determine the length of the transverse waves created by the periscope.

15. A 20-ft-deep stream flows past a 1-ft-diameter pile. The current velocity is 10 fps.
   (a) Determine the angle of the wave envelope.
   (b) If $A(\theta) = 1$, determine the wave drag on the pile, assuming deep water, where $\gamma = 62.4$ lb/ft^3.
   (c) what are the magnitudes of the transverse and divergent wave-induced forces on the pile?

Chapter

FOUR

FLOATING STRUCTURES IN WAVES

A basic requirement in the design of a floating system such as a ship, working platform, or buoy is that the designer understand the nature of the forces on and the motions of a body floating in a sea. A lack of this understanding can result in an underdesigned system that may experience either structural or mission failure. For example, a ship in a heavy sea may fracture amidships because of severe wave-induced bending moments in the hull. A floating platform may experience such severe motions that performance of the task for which it was designed is not possible.

This chapter is devoted to the mathematical and physical descriptions of the coupled heaving and pitching motions of bodies in regular seas. In the first part of the chapter freely floating bodies are discussed; in the second part the forces on and the motions of moored bodies are described.

4.1. COUPLED HEAVING AND PITCHING

The motions experienced by a floating body are the rectilinear motions of *heave, surge, and sway* and the angular motions of *pitch, roll, and yaw*, as illustrated in Figure 4.1. Since there are six degrees of

4.1. COUPLED HEAVING AND PITCHING

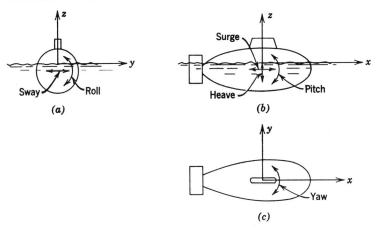

Figure 4.1. Six degrees of freedom of a floating body. (*a*) Front view, (*b*) Side view, (*c*) Top view.

freedom, a complete analysis of the motion of the body requires the solution of six simultaneous equations. In this section only the motions of heave and pitch are described. With a complete understanding of these planar motions the reader can easily comprehend the analyses of the remaining four degrees of freedom as described in the book by Blagoveshchensky (1962) and in the volume edited by Comstock (1967).

The method of analysis used in this section is known as the *strip theory*. The basic assumption of the theory is that the time-dependent flow adjacent to the wet surfaces of thin vertical slices of the floating body is two-dimensional, that is, the fore-and-aft components of the flow are considered to be of second order. It is then assumed that the resulting forces and moments on these elemental slices or strips can be summed over the length of the body to obtain the total forces and moments. Furthermore, these two assumptions are coupled with the *Froude-Krylov hypothesis*, which states that the motions of a body do not alter particle motions in the surface wave, although the particle motions influence the motions of the body.

A freely floating body such as that shown in Figure 4.2 is subject to the following forces: the *inertial reaction of the fluid*, F_1; the *hydrostatic restoring force*, F_2; a *damping force*, F_3; and a *force induced by surface*

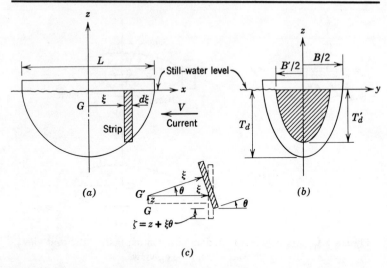

Figure 4.2. The strip of a floating body. (*a*) Side view. (*b*) Front view. (*c*) Displacement of the strip.

waves, F_4. The force F_4 is actually a special case of the inertial and hydrostatic forces and acts as the *forcing function* of the system. Applying Newton's second law of motion to the vertical motions of the body results in the following expression for the *heaving*:

$$m \frac{d^2 z}{dt^2} = F_1 + F_2 + F_3 + F_4 \qquad (4.1)$$

In addition to the rectilinear heaving motion, these forces cause an angular *pitching motion*, described by

$$I_y \frac{d^2 \theta}{dt^2} = M_1 + M_2 + M_3 + M_4 \qquad (4.2)$$

where I is the moment of inertia, and M_j is the pitching moment due to the force F_j.

Before describing the forces and moments appearing in Eqs. 4.1 and 4.2 the motion of the strip must be discussed. Consider the body sketched in Figure 4.2 to be symmetric about both the vertical $x - z$ and $y - z$ planes. This symmetry is characteristic of most buoy shapes. First, it is assumed that the body is in a calm sea with a current of velocity **V**. The displacement of the body from its equilibrium

4.1. COUPLED HEAVING AND PITCHING

position is assumed to be small enough to allow the linearizing approximation

$$\theta \simeq \sin(\theta) \simeq \tan(\theta) \quad (4.3)$$

The thin section of the body, called the *strip*, is located at a horizontal distance ξ from the center of gravity, G. The forces on the strip are assumed to be only in the $y - z$ plane. In addition, the wetted surface of the strip is assumed to have no curvature in the x (or ξ) direction. This last assumption becomes somewhat questionable, however, when the body in question has significant curvature. For this reason the strip theory is most applicable to bodies having ship shapes.

When the body rises by an amount z while rotating through an angle θ, as illustrated in Figure 4.2c, the total vertical displacement of the strip is

$$\zeta = z + \xi\theta \quad (4.4)$$

where the approximation of Eq. 4.3 is used. If the vertical displacement is time dependent, then the vertical velocity is

$$\dot{\zeta} = \dot{z} + \xi\dot{\theta} + \dot{\xi}\theta$$

where the dot notation is used to represent time derivatives. The term $\dot{\xi}\theta$ is the vertical component of the velocity due to the instantaneous *trim angle* θ. Since the motion of the body is relative to the fluid, one can see that the relative velocity between the body and the fluid is

$$w_b = \dot{z} + \xi\dot{\theta} + \dot{\xi}\theta - V\theta \quad (4.5)$$

where, again, the velocity $V\theta$ is due to the trim angle θ. To gain a physical understanding of the component $V\theta$ consider the velocity diagram shown in Figure 4.3. The water particles strike the surface, which is inclined at an angle θ. Thus the tangential component of the current velocity is

$$v_t = V\cos(\theta) \simeq V$$

and the vertical component of v_t is

$$v_t \sin(\theta) \simeq v_t\theta \simeq V\theta$$

The vertical acceleration of the strip is

$$\dot{w}_b = \ddot{z} + \xi\ddot{\theta} + (2\dot{\xi} - V)\dot{\theta} + \ddot{\xi}\theta \quad (4.6)$$

116 FLOATING STRUCTURES IN WAVES

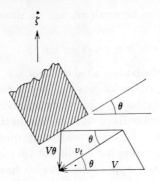

Figure 4.3. Velocity diagram at bottom of the strip.

where the current velocity V is constant, so that $\dot{V} = 0$. The velocity $\dot{\xi}$ and the acceleration $\ddot{\xi}$ are relatively small and can be neglected without serious error. This assumption is made throughout the rest of the discussion.

A. Motion-Induced Force and Moment

Since the fluid adjacent to the strip is accelerating with the strip, the inertial reaction of the fluid acts as the following external force on the strip:

$$\frac{dF_1}{d\xi} = -\frac{D}{Dt}[m_w(\xi)w_b]$$

$$= -\frac{\partial}{\partial t}[m_w(\xi)w_b] -$$

$$(-V\mathbf{i} + v\mathbf{j} + w\mathbf{k}) \cdot \left(\frac{\partial}{\partial \xi}\mathbf{i} + \frac{\partial}{\partial y}\mathbf{j} + \frac{\partial}{\partial z}\mathbf{k}\right)[m_w(\dot{z} + \xi\dot{\theta} - V\theta)]$$

$$= -m_w(\ddot{z} + \xi\ddot{\theta} - 2V\dot{\theta}) + V\frac{dm_w}{\partial \xi}(\dot{z} + \xi\dot{\theta} - V\theta) \qquad (4.7)$$

where $m_w(\xi)$ is the *added-mass per unit length* for the particular cross-

Plate 4.1. The deep submergence vehicle ALVIN experiencing rather heavy seas. (Courtesy of the Woods Hole Oceanographic Institution.)

section of the body, as described in Section 3.5,* and where $\dot{\xi} \simeq \ddot{\xi} \simeq 0$. The theoretical derivations of the expressions of added-mass for various cross sections are beyond the scope of this book. Some expressions for simple cross sections derived by Lamb (1945), Lewis (1929), and Wendel (1956) are presented in Table 3.1.

The moment about the center of gravity of the body resulting from the motion-induced force is

$$\frac{dM_1)^+}{d\xi} = \xi \frac{dF_1}{d\xi} \qquad (4.8)$$

Integrating Eqs. 4.7 and 4.8, respectively, over the length of the body yields the *total motion-induced force and moment* on the body, that is,

$$F_1 = \int_{-L/2}^{L/2} \frac{dF_1}{d\xi} d\xi$$

$$= -\int_{-L/2}^{L/2} m_w\, d\xi\, \ddot{z} - \int_{-L/2}^{L/2} \xi m_w\, d\xi\, \ddot{\theta} + V \int_{-L/2}^{L/2} \left(\xi \frac{dm_w}{d\xi} + 2m_w\right) d\xi\, \dot{\theta} \qquad (4.9)$$

since

$$\int_{-L/2}^{L/2} \frac{dm_w}{d\xi}\, d\xi = 0$$

and

$$M_1)^+ = \int_{-L/2}^{L/2} \xi \frac{dF_1}{d\xi}\, d\xi$$

$$= -\int_{-L/2}^{L/2} \xi m_w\, d\xi\, \ddot{z} - \int_{-L/2}^{L/2} \xi^2 m_w\, d\xi\, \ddot{\theta} + V \int_{-L/2}^{L/2} \xi \frac{dm_w}{d\xi}\, d\xi\, \dot{z}$$

$$+ V \int_{-L/2}^{L/2} \left(\xi^2 \frac{dm_w}{d\xi} + 2\xi m_w\right) d\xi\, \dot{\theta} - V^2 \int_{-L/2}^{L/2} \xi \frac{dm_w}{d\xi}\, d\xi\, \theta \qquad (4.10)$$

* For half-submerged bodies having cross sections that are shown in Table 3.1, half the added-mass value presented in the table should be used for the corresponding cross section. This approximation is satisfactory for the purposes of this book; however, the reader should consult the report by Wendel (1956) and the section by Lewis in the book edited by Comstock (1967) for more thorough and accurate treatments of the added-mass and moment of inertia of floating bodies.

4.1. COUPLED HEAVING AND PITCHING

To illustrate the use of Eqs. 4.9 and 4.10, consider a sphere floating half-submerged, as sketched in Figure 4.4. The strip shown in the figure has the shape of a semicylindrical disk; therefore, from Table 3.1, the expression for the added-mass of the strip is

$$m_w(\xi) = \tfrac{1}{2}\rho\pi\left(\frac{B'}{2}\right)^2$$

where, referring to Figure 4.4c,

$$B'(\xi) = 2\sqrt{(L/2)^2 - \xi^2}$$

Thus

$$m_w = \tfrac{1}{2}\rho\pi\left[\left(\frac{L}{2}\right)^2 - \xi^2\right]$$

and

$$\frac{dm_w}{d\xi} = -\rho\pi\xi$$

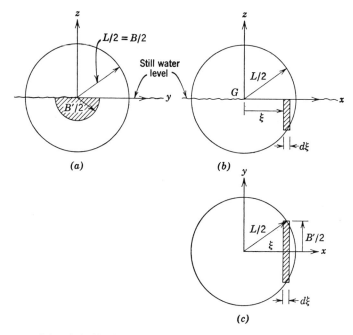

Figure 4.4. A half-submerged floating sphere. (a) Front view. (b) Side view. (c) Top view.

The motion-induced force on the sphere, from Eq. 4.9, is then

$$F_1 = -\tfrac{1}{12}\pi\rho L^3 \ddot{z} + \tfrac{1}{12}V\pi\rho L^3 \dot{\theta}$$

Applying Eq. 4.10 to the spherical float results in the following expression for the motion-induced moment:

$$M_1)^+ = -\tfrac{1}{12}\rho\pi L^3 \left[\left(\frac{L^2}{20}\right)\ddot{\theta} + V\dot{z} - V^2\theta\right]$$

B. Hydrostatic Restoring Force and Moment

When the strip is displaced from its position of equilibrium as shown in Figure 4.2c, there is a change in the *buoyant force* due to the gain or loss of displaced volume. Since the change in displacement shown in Figure 4.2c is negative, the resulting hydrostatic restoring force on the strip is

$$\frac{dF_2}{d\xi} = -\rho g B'(\xi)\zeta \qquad (4.11)$$

The use of Eq. 4.11 requires the assumption that the body is "*wall-sided*" at the water line, that is, there is no curvature in the $y - z$ plane about the free surface. This assumption is valid for bodies having moderate curvature at the water line but experiencing small displacements. The moment due to the force in Eq. 4.11 is

$$\frac{dM_2)^+}{d\xi} = \xi \frac{dF_2}{d\xi} \qquad (4.12)$$

The *total hydrostatic restoring force and moment* are obtained by integrating Eqs. 4.11 and 4.12, respectively, over the length of the body. Thus, using the results in Eq. 4.4, one obtains

$$F_2 = -\rho g \int_{-L/2}^{L/2} B'(\xi)\,d\xi\, z - \rho g \int_{-L/2}^{L/2} \xi B'(\xi)\,d\xi\,\theta \qquad (4.13)$$

and

$$M_2)^+ = \int_{-L/2}^{L/2} \xi \frac{dF_2}{d\xi}\,d\xi$$

$$= -\rho g \int_{-L/2}^{L/2} \xi B'(\xi)\,d\xi\, z - \rho g \int_{-L/2}^{L/2} \xi^2 B'(\xi)\,d\xi\,\theta \qquad (4.14)$$

4.1. COUPLED HEAVING AND PITCHING

Applying the results of Eqs. 4.13 and 4.14 to the spherical float shown in Figure 4.4 results in the following expressions:

$$F_2 = -\frac{\pi \rho g L^2 z}{4}$$

and

$$M_2)^+ = -\tfrac{1}{64}\pi \rho g L^4 \theta$$

respectively.

C. Damping

The *damping* of the motion of a floating body is due to both viscosity and energy lost in the *creation of waves*. In this section only the latter is discussed since viscous damping is assumed to be of second order for slight motions.

Consider a strip of the body shown in Figure 4.2 to be oscillating in the vertical plane, as shown in Figure 4.5a. The *radiated surface waves* are created on both sides of the strip because of an inclination of the sides at the water line, adhesion of the water to the hull, or a change in the displacement of the body. A basic assumption of the strip theory is that the body is wall-sided at the water line, that is, there is no curvature of the surface of the body at the free surface. Furthermore, the fluid is assumed to be inviscid and to have no surface tension; therefore the damping waves are assumed to be caused by the alternating volume of displacement caused by the rise and fall of the body.

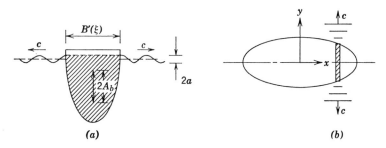

Figure 4.5. Damping wave generation. (*a*) Vertically oscillating strip. (*b*) Radiation of surface waves.

The power transmitted to the wave (per cycle) is represented by the energy flux of the wave as expressed by Eq. 2.47;

$$\dot{E}\,d\xi = \frac{\rho g a_d^2 c_g}{2}\,d\xi \qquad (4.15)$$

where c_g is the group velocity described in Eq. 2.41. In the present analysis the assumption of deep water is made so that

$$c_g \to \frac{c}{2} = \frac{\omega_n}{2k}$$

and

$$k \to \frac{\omega_n^2}{g}$$

from Eqs 2.41 and 2.40, respectively. Here ω_n is the frequency of the body that is equal to the wave frequency when the body is excited by surface waves. The power transmitted in deep water is then

$$\dot{E}\,d\xi = \frac{\rho g^2 a_d^2}{4\omega_n}\,d\xi \qquad (4.16)$$

The power lost to the water is equal to the product of the damping force, dF_d, and the vertical velocity of the body; that is, if $A_b(\xi)$ is the amplitude of the body motion at a position ξ, then over one wave period the power lost by the body is

$$dF_d\,\bar{w}_b = dF_d\left(\frac{4A_b}{T}\right) = dF_d\left(\frac{2A_b\,\omega_n}{\pi}\right) \qquad (4.17)$$

where T is the period of both the body and the wave motion, and $\bar{w}_b$ is the average speed. The amplitudes of the wave and body motion need not be equal.

Since the power lost by the body must be equal to that gained by the water, the expressions in Eqs. 4.16 and 4.17 can be equated to obtain

$$dF_d\,\bar{w}_b = 2\dot{E}\,d\xi \qquad (4.18)$$

4.1. COUPLED HEAVING AND PITCHING

where the factor 2 occurs because of the two sets of waves being generated. By combining Eqs. 4.15, 4.16, and 4.17 and simplifying the result, the following expression for the damping force per unit body length is obtained:

$$\frac{dF_d}{d\xi} = \frac{2\dot{E}}{\bar{w}_b} = \frac{\pi \rho g^2 A_b (a_b/A_b)^2}{4\omega_n^2} \quad (4.19)$$

Since the force is dependent on the vertical velocity, dividing Eq. 4.19 by the average vertical speed, $\bar{w}_b = 4A/T$, yields the *damping force per unit vertical velocity of the* strip:

$$\frac{dF_d}{d\xi} \frac{1}{\bar{w}_b} = \frac{dN}{d\xi} = \frac{\pi^2 \rho g^2 (a_d/A_b)^2}{8\omega_n^3} \quad (4.20)$$

where $N(\xi)$ is the damping force per unit vertical velocity. The damping force on the strip is now

$$\frac{dF_3}{d\xi} = -\frac{dN}{d\xi} w_b \quad (4.21)$$

and the resulting moment about the center of gravity of the floating body is

$$\frac{dM_3}{d\xi}\bigg)^+ = \xi \frac{dF_3}{d\xi} = -\xi \frac{dN}{d\xi} w_b \quad (4.22)$$

The *total damping force* and *moment* on the body are obtained by integrating Eqs. 4.21 and 4.22 over the length L. The respective results are

$$F_3 = -\int_{-L/2}^{L/2} \left(\frac{dN}{d\xi}\right) w_b \, d\xi$$

$$= -\frac{\pi^2 \rho g^2}{8\omega_n^3} \int_{-L/2}^{L/2} \bar{A}^2(\xi) w_b \, d\xi \quad (4.23)$$

and

$$M_3\bigg)^+ = -\int_{-L/2}^{L/2} \xi \left(\frac{dN}{d\xi}\right) w_b \, d\xi$$

$$= -\frac{\pi^2 \rho g^2}{8\omega_n^3} \int_{-L/2}^{L/2} \xi \bar{A}^2(\xi) w_b \, d\xi \quad (4.24)$$

where $\bar{A}(\xi)$ is the ratio of the wave amplitude and the amplitude of motion of the strip, that is, $\bar{A}(\xi) = a_d/A_b$.

It should be noted that the damping analysis presented here is two dimensional in that the waves generated by each strip travel only in the y direction, as illustrated in Figure 4.5b. Obviously, this is a simplification since one can observe waves generated in all directions from a disturbance in quiet water. Furthermore, the determination of the values of the amplitude ratio $\bar{A}(\xi)$ for various cross sections requires considerably more effort than can (or should) be devoted to this topic in a book of this type. The theoretical derivation of the term $\bar{A}(\xi)$ was performed by Grim (1959). Data obtained from Grim's theory can be found in the report by Jacobs et al. (1960) and in the volume edited by Comstock (1967). For the purposes of this book $\bar{A}(\xi)$ is considered to be a constant that must be determined experimentally. By treating $\bar{A}(\xi)$ in this manner the errors resulting from the previously mentioned simplifications are minimized.

Now, if $\bar{A}$ is assumed to be independent of position, the total damping force and moment on any floating body obtained from Eqs. 4.23 and 4.24, respectively, are

$$F_3 = -\frac{\pi^2 \rho g^2}{8\omega_n^3} \bar{A}^2 L(\dot{z} - V\theta) \qquad (4.25)$$

and

$$M_3)^+ = -\frac{\pi^2 \rho g^2}{96\omega_n^3} \bar{A}^2 L^3 \dot{\theta} \qquad (4.26)$$

An illustration of the method of determining the value of $\bar{A}$ is given in Section 4.3.

D. Wave-Induced Force and Moment

Surface waves can be considered as the *forcing function* of a floating system. The forces acting on a floating body that are induced by waves are both inertial and hydrostatic in nature, however, these forces differ from those described in Sections 4.1A and 4.1B in that the time dependence is explicit in the expressions for the wave-induced forces, as will be shown.

4.1. COUPLED HEAVING AND PITCHING

First, consider the inertial reaction of the water particles within a wave passing a hull. As previously mentioned, the Froude-Krylov hypothesis is assumed to hold, that is, the presence of the floating body does not alter the pressure field in the wave from that existing when no body is present. The inertial reaction of the fluid can be analyzed by considering the hydrodynamic pressure at any point within the wave. If the flow is assumed to be irrotational, this pressure is described by Eq. 2.46, where the velocity potential is given in Eq. 2.22. Combining these equations and applying the result to deep water yields

$$p_w = -\rho \frac{\partial \phi}{\partial t} = \rho a_w g e^{kz} \cos(kx - \omega t) \tag{4.27}$$

In the case of Figure 4.6, the pressure force on an elemental area of a strip is

$$p_w \, ds \, d\xi = -\rho \frac{\partial \phi}{\partial t} \, ds \, d\xi$$

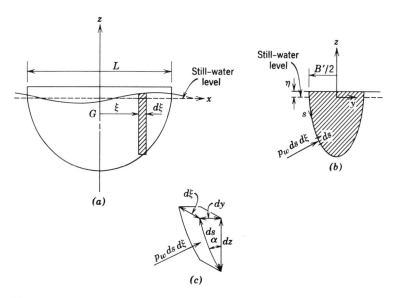

Figure 4.6. A floating body in waves. (a) Side view. (b) The strip. (c) Elemental pressure force.

where s is the curvilinear coordinate along the wetted perimeter of the strip, as shown in Figure 4.6b. Since

$$s = s(y, z)$$

then

$$ds = dy \sin(\alpha) + dz \cos(\alpha)$$

from Figure 4.6c. The vertical component of the pressure force is then

$$p_w \, ds \quad d\xi \quad \sin(\alpha) = p_w \, (dy) \, d\xi$$

where $dy/dz = \tan(\alpha)$. Thus the vertical component of the wave-induced hydrodynamic pressure force acting on the strip is

$$\frac{dF_{4a}}{d\xi} = \int_{-B'/2}^{B'/2} p_w \, dy$$

$$= \rho a_w g \cos(k\xi - \omega t) \int_{-B'/2}^{B'/2} e^{kz} \, dy \qquad (4.28)$$

where the integration is about the wetted perimeter, that is, $z = z(y)$.

The second wave-induced force is the change in the hydrostatic restoring force due to the passage of a wave. Using the linear-wave free-surface displacement described in Eq. 2.20, one can express the buoyant force on the strip as

$$\frac{dF_{4b}}{d\xi} = \rho g B' \eta$$

$$= \rho g B' a \cos(k\xi - \omega t) \qquad (4.29)$$

where the breadth of the strip is a function of position, that is, $B' = B'(\xi)$.

The total vertical wave-induced force on the strip is the sum of the expressions in Eqs. 4.28 and 4.29:

$$\frac{dF_4}{d\xi} = \frac{d}{d\xi}(F_{4a} + F_{4b})$$

$$= \rho g a_w \cos(k\xi - \omega t)\left(\int_{-B'/2}^{B'/2} e^{kz} \, dy + B'\right)$$

$$= \rho g a_w \left(\int_{-B'/2}^{B'/2} e^{kz} + B'\right)[\cos(k\xi)\cos(\omega t) + \sin(k\xi)\sin(\omega t)]$$

$$(4.30)$$

4.1. COUPLED HEAVING AND PITCHING

The moment about the center of gravity of the body resulting from the force described in Eq. 4.30 is

$$\frac{dM_4)^+}{d\xi} = \xi \frac{dF_4}{d\xi} \tag{4.31}$$

The *total wave-induced force and moment* on a floating body is obtained by integrating Eqs. 4.30 and 4.31, respectively, over the wetted hull length:

$$F_4 = \rho g a_w \int_{-L/2}^{L/2} \left[\int_{-B'/2}^{B'/2} e^{kz} \, dy + B'(\xi) \right] \cdot$$
$$[\cos(k\xi)\cos(\omega t) + \sin(k\xi)\sin(\omega t)] \, d\xi \tag{4.32}$$

and

$$M_4)^+ = \rho g a_w \int_{-L/2}^{L/2} \xi \left[\int_{-B'/2}^{B'/2} e^{kz} \, dy + B'(\xi) \right] \cdot$$
$$[\cos(k\xi)\cos(\omega t) + \sin(k\xi)\sin(\omega t)] \, d\xi \tag{4.33}$$

By comparing the expressions in Eqs. 4.32 and 4.33 with the expressions for the forces and moments contained in Sections 4.1A, 4.1B, and 4.1C, one can readily see that the time dependence is explicit only in the wave-induced force and moment equations; therefore F_4 and M_4 are the *exciting force* and *moment* of the dynamic system.

To apply the results in Eqs. 4.32 and 4.33 to a floating body of any sectional area it is necessary, first, to describe the mathematical form of the sectional perimeter, that is, $z = z(y)$. As an example, again consider the floating sphere sketched in Figure 4.4. The expression for the perimeter of the semicircular section shown in Figure 4.7 is

$$z = -\left(\frac{B'}{2}\right)\sin(\alpha) = y\tan(\alpha)$$

where

$$B' = 2\sqrt{(L/2)^2 - \xi^2}$$

Hence the y integral in Eqs. 4.32 and 4.33 becomes

$$\int_{-B'/2}^{B'/2} e^{kz} \, dy = \int_0^\pi \left(\frac{B'}{2}\right) e^{-k(B'/2)\sin(\alpha)} \sin(\alpha) \, d\alpha$$

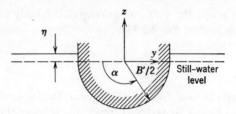

Figure 4.7. Strip of a floating sphere in waves.

The integration is accomplished by, first, expanding the exponential term of the second integrand into the following infinite series:

$$e^{-k(B'/2)\sin(\alpha)} = 1 - \left(\frac{kB'}{2}\right)\sin(\alpha) + \frac{1}{2!}\left(\frac{kB'}{2}\right)^2 \sin^2(\alpha) \cdots$$

The integral then becomes

$$\int_0^\pi \left(\frac{B'}{2}\right) e^{-k(B'/2)\sin(\alpha)} \sin(\alpha)\, d\alpha = B' - \frac{\pi}{8} kB'^2 + \frac{k^2 B'^3}{12} \cdots \quad (4.34)$$

Because of the mathematical form of $B' = B'(\xi)$, the combinations of Eq. 4.34 and the force and moment expressions of Eqs. 4.32 and 4.33, respectively, result in rather unwieldy integrals. To avoid becoming overwhelmed by integration techniques, the assumption is made that the waves being encountered by the floating sphere are of such length as to cause the terms involving powers of the wave number, k, of 2 and higher to be negligible. In other words, the wavelength is assumed to be much longer than the maximum breadth of the body. With this assumption, Eqs. 4.32 and 4.33 become

$$F_4 = \rho g a_w \int_{-L/2}^{L/2} \left(2B' - \frac{\pi}{8} kB'^2\right)[\cos(k\xi)\cos(\omega t) +$$
$$\sin(k\xi)\sin(\omega t)]\, d\xi \quad (4.35)$$

and

$$M_4\}^+ = \rho g a_w \int_{-L/2}^{L/2} \xi\left(2B' - \frac{\pi}{8} kB'^2\right)[\cos(k\xi)\cos(\omega t) +$$
$$\sin(k\xi)\sin(\omega t)]\, d\xi \quad (4.36)$$

respectively.

4.1. COUPLED HEAVING AND PITCHING

The integrations involved in Eqs. 4.35 and 4.36 must still be performed by using the series technique since the integrals of $B' \cos(k\xi)$ and $B' \sin(k\xi)$ have no closed solutions. Hence the trigonometric terms involving ξ are expanded into the following infinite series:

$$\cos(k\xi) = 1 - \frac{1}{2!}(k\xi)^2 + \frac{1}{4!}(k\xi)^4 \cdots$$

and

$$\sin(k\xi) = (k\xi) - \frac{1}{3!}(k\xi)^3 \cdots$$

Since the long-wave assumption makes it possible to neglect the terms involving k^n where $n \geq 2$, the wave-induced force and moment on a floating sphere that is experiencing long waves in deep water are, respectively,

$$F_4 = \frac{\pi \rho g a_w L^2}{2}\left(1 - \frac{kL}{6}\right) \cos(\omega t)$$

and

$$M_4)^+ = \frac{\pi \rho g a_w k L^4}{32} \sin(\omega t)$$

From these results, it is evident that the wave-induced force and moment are out of time phase by $\pi/2$, that is, the force leads the moment by $\pi/2$.

E. Equations of Motion of Heave and Pitch

In Sections 4.1A through 4.1D the forces and moments experienced by a floating body were described. With the exception of the wave-induced forces and moments presented in Section 4.1D, the results were found to depend on the rectilinear and rotational motions of the body. The wave-induced force and moment, however, are independent of the body motions (Froude-Krylov hypothesis) and act as the forcing function of the floating system. With these results, the equations of motion of the body are now formed.

Equations 4.9, 4.13, 4.23, and 4.32 are combined with Eq. 4.1 to obtain the equation of heaving motion:

$$m\ddot{z} + a\ddot{z} + b\dot{z} + cz + d\ddot{\theta} + e\dot{\theta} + h\theta$$
$$= F_A \cos(\omega t) + F_B \sin(\omega t) \quad (4.37)$$
$$= F_0 \cos(\omega t + \gamma)$$

where the hydrodynamic coefficients are

$$\left.\begin{aligned}
a &= \int_{-L/2}^{L/2} m_w \, d\xi \\
b &= \int_{-L/2}^{L/2} \frac{dN}{d\xi} d\xi = \frac{\pi^2 \rho g^2}{8\omega_n^3} \int_{-L/2}^{L/2} \bar{A}^2 \, d\xi \\
c &= \rho g \int_{-L/2}^{L/2} B' \, d\xi \\
d &= \int_{-L/2}^{L/2} \xi m_w \, d\xi \\
e &= -V \int_{-L/2}^{L/2} \left(\xi \frac{dm_w}{d\xi} + 2m_w\right) d\xi + \frac{\pi^2 \rho g^2}{8\omega_n^3} \int_{-L/2}^{L/2} \xi \bar{A}^2 \, d\xi \\
h &= \rho g \int_{-L/2}^{L/2} \xi B' \, d\xi - \frac{V \pi^2 \rho g^2}{8\omega_n^3} \int_{-L/2}^{L/2} \bar{A}^2 \, d\xi \\
F_A &= \rho g a_w \int_{-L/2}^{L/2} \left(\int_{-B'/2}^{B'/2} e^{kz} dy + B'\right) \cos(k\xi) \, d\xi \\
F_B &= \rho g a_w \int_{-L/2}^{L/2} \left(\int_{-B'/2}^{B'/2} e^{kz} dy + B'\right) \sin(k\xi) \, d\xi \\
F_0 &= \sqrt{(F_A^2 + F_B^2)}
\end{aligned}\right\} \quad (4.38)$$

and

$$\gamma = \tan^{-1}\left(\frac{-F_B}{F_A}\right)$$

Similarly, the equation describing the *pitching motion* is obtained by combining Eqs. 4.10, 4.14, 4.24, and 4.33 with 4.2:

4.1. COUPLED HEAVING AND PITCHING

$$I_y \ddot{\theta} + A\ddot{\theta} + B\dot{\theta} + C\theta + D\ddot{z} + E\dot{z} + Hz$$
$$= M_A \cos(\omega t) + M_B \sin(\omega t) \quad (4.39)$$
$$= M_0 \cos(\omega t + \delta)$$

where the coefficients are the following:

$$\left. \begin{aligned} A &= \int_{-L/2}^{L/2} \xi^2 m_w \, d\xi \\ B &= -V \int_{-L/2}^{L/2} \xi \left(\xi \frac{dm_w}{d\xi} + 2m_w \right) d\xi + \frac{\pi^2 \rho g^2}{8\omega_n^3} \int_{-L/2}^{L/2} \xi^2 \bar{A}^2 \, d\xi \\ C &= V^2 \int_{-L/2}^{L/2} \xi \frac{dm_w}{d\xi} d\xi + \rho g \int_{-L/2}^{L/2} \xi^2 B' \, d\xi - \frac{V\pi^2 \rho g^2}{8\omega_n^3} \int_{-L/2}^{L/2} \xi \bar{A}^2 \, d\xi \\ D &= d \\ E &= -V \int_{-L/2}^{L/2} \xi \frac{dm_w}{d\xi} d\xi + \frac{\pi^2 \rho g^2}{8\omega_n^3} \int_{-L/2}^{L/2} \xi \bar{A}^2 \, d\xi \\ H &= \rho g \int_{-L/2}^{L/2} \xi B' \, d\xi \\ M_A &= \rho g a_w \int_{-L/2}^{L/2} \xi \left(\int_{-B'/2}^{B'/2} e^{kz} \, dy + B' \right) \cos(k\xi) \, d\xi \\ M_B &= \rho g a_w \int_{-L/2}^{L/2} \xi \left(\int_{-B'/2}^{B'/2} e^{kz} \, dy + B' \right) \sin(k\xi) \, d\xi \\ M_0 &= \sqrt{(M_A^2 + M_B^2)} \end{aligned} \right\} \quad (4.40)$$

and

$$\delta = \tan^{-1}\left(\frac{-M_B}{M_A}\right)$$

Before discussing the solution of the system of equations that describe the coupled heaving and pitching motions of a floating body, it is helpful to consider Eqs. 4.37 and 4.39 as applied to the spherical float shown in Figure 4.4. The coefficients in these equations were derived in Sections 4.1A through 4.1D. Using these results, and assuming $\bar{A}$ to be constant, one can write the heaving equation as

$$\left(m + \frac{\pi \rho L^3}{12}\right)\ddot{z} + \frac{\pi^2 \rho g^2 \bar{A}^2 L}{8\omega_n^3}\dot{z} + \frac{\pi \rho g L^2}{4} z - \frac{V\pi \rho L^3}{12}\dot{\theta} -$$

$$\frac{V\pi^2 \rho g^2 \bar{A}^2 L}{8\omega_n^3}\theta = \frac{\pi \rho g a_w L^2}{2}\left(1 - \frac{kL}{6}\right)\cos(\omega t)$$

and the pitching equation as

$$\left(I_y + \frac{\pi \rho L^5}{240}\right)\ddot{\theta} + \frac{\pi^2 \rho g^2 \bar{A}^2 L^3}{96\omega_n^3}\dot{\theta} + \left(\frac{\pi \rho g L^4}{64} - \frac{V^2 \pi \rho L^3}{12}\right)\theta +$$

$$\frac{V\pi \rho L^3}{12}\dot{z} = \frac{\pi \rho g a_w k L^4}{32}\left(1 - \frac{2}{15}kL\right)\sin(\omega t)$$

Thus, for the half-submerged sphere, $d = D = H = 0$. When these coefficients are examined, one sees that the reason for their zero values is the symmetry of the body about the $y - z$ plane.

From the results obtained for the sphere and from a further examination of the coefficients described in Eqs. 4.38 and 4.40, six generalizations can be made.

CASE 1

For bodies having *symmetry* about the $y - z$ plane the following results are obtained:

$$\int_{-L/2}^{L/2} \xi m_w \, d\xi = \int_{-L/2}^{L/2} \xi B' \, d\xi = \int_{-L/2}^{L/2} \xi^2 \frac{dm_w}{d\xi} \, d\xi = \int_{L/2}^{L/2} \xi \bar{A}^2 \, d\xi = 0$$

CASE 2

If the ratio of the *damping* wave amplitude and the amplitude of the vertical body motion, that is, $\bar{A} = a_d/A_b$, is assumed to be constant, then

$$\int_{-L/2}^{L/2} \xi \bar{A}^2 \, d\xi = \bar{A}^2 \int_{-L/2}^{L/2} \xi \, d\xi = 0$$

CASE 3

When *no current* exists or if the body has no forward motion, then $e = E = 0$, assuming body symmetry.

4.1. COUPLED HEAVING AND PITCHING

CASE 4

For a *fully submerged* body within the region of influence of the free surface, $c = H = 0$ and, also, any term having B' in the integrand is zero except in the limits of the y-integration for F_A, F_B, M_A, and M_B.

CASE 5

For a *spar buoy*, that is, one having a significant depth-to-breadth ratio, extending below the region of free-surface influence, that is, below a depth of $\lambda/2$, the y integrals in the expressions for F_A, F_B, M_A, and M_B are all equal to zero.

CASE 6

A *symmetric* body for which $\bar{A}$ is *constant* and $V = 0$ undergoes heaving and pitching motions that are independent of each other since $d = e = h = D = E = H = 0$. In this case the equations of motion are said to be *uncoupled*.

Returning to the equations of motion, Eqs. 4.37 and 4.39, one writes these equations in *operator* form, so that the heaving equation is

$$L_{11} z + L_{12} \theta = F(t) \tag{4.41}$$

and the pitching equation is

$$L_{21} z + L_{22} \theta = M(t) \tag{4.42}$$

where the L operators are defined as follows:

$$\left. \begin{aligned} L_{11} &= (m + a) \frac{d^2}{dt^2} + b \frac{d}{dt} + c \\ L_{12} &= d \frac{d^2}{dt^2} + e \frac{d}{dt} + h \\ L_{21} &= D \frac{d^2}{dt^2} + E \frac{d}{dt} + H \\ L_{22} &= (Iy + A) \frac{d^2}{dt^2} + B \frac{d}{dt} + C \end{aligned} \right\} \tag{4.43}$$

and

Now, applying *Cramer's rule* for linear equations, as described by Wylie (1960), to Eqs. 4.41 and 4.42, one obtains the following equations in z and θ, respectively:

$$\begin{vmatrix} L_{11} & L_{12} \\ L_{21} & L_{22} \end{vmatrix} z = \begin{vmatrix} F(t) & L_{12} \\ M(t) & L_{22} \end{vmatrix} \quad (4.44a)$$

and

$$\begin{vmatrix} L_{11} & L_{12} \\ L_{21} & L_{22} \end{vmatrix} \theta = \begin{vmatrix} L_{11} & F(t) \\ L_{21} & M(t) \end{vmatrix} \quad (4.45a)$$

or, in expanded form, the heaving equation is

$$(L_{11}L_{22} - L_{12}L_{21})z = L_{22}F(t) - L_{12}M(t) \quad (4.44b)$$

and the pitching equation is

$$(L_{11}L_{22} - L_{12}L_{21})\theta = L_{11}F(t) - L_{21}M(t) \quad (4.45b)$$

Equations 4.44 and 4.45 are fourth-order, inhomogeneous, linear, differential equations that can be solved using standard techniques, as described by Ince (1956).

To obtain an idea of the significance of the terms in the equations of motion consider the motions of a floating symmetric body subject to deep-water waves with no current, assuming the damping ratio $\bar{A}$ to be constant. This is the situation described in Case 6; hence the equations of motion are uncoupled. Equations 4.37 and 4.39 reduce to

$$(m + a)\ddot{z} + b\dot{z} + cz = F_0 \cos(\omega t + \gamma) \quad (4.46)$$

and

$$(I_y + A)\ddot{\theta} + B\dot{\theta} + C\theta = M_0 \cos(\omega t + \delta) \quad (4.47)$$

respectively. The solution of either equation is obtained by following the same procedure. Therefore attention can be confined to the heaving equation (Eq. 4.46). The solution of this equation consists of a *transient part* and a *steady-state part*. The steady-state solution is of significance here and can be written as

$$\begin{aligned} z &= Z_0 \cos(\omega t + \gamma - \sigma) \\ &= \frac{F_0/c}{\sqrt{(1 - \omega^2/\omega_n^2)^2 + [2\Delta\omega/\omega_n]^2}} \cos(\omega t + \gamma - \delta) \end{aligned} \quad (4.48)$$

4.1. COUPLED HEAVING AND PITCHING

where the phase angle σ is

$$\sigma = \tan^{-1}\left[\frac{2\Delta\omega/\omega_n}{1-(\omega/\omega_n)^2}\right] \tag{4.49}$$

and is shown in Figure 4.8. The term

$$\Delta = \frac{b}{2\sqrt{c(m+a)}} \tag{4.50}$$

is the ratio of the damping term, b, and the *critical damping*, $2\sqrt{c(m+a)}$, and

$$\omega_n = \sqrt{c/(m+a)} \tag{4.51}$$

is the *natural frequency* of the body in heave. The term F_0/c is the *static displacement* of the body, and the ratio

$$Z = \frac{Z_0}{F_0/c} \tag{4.52}$$

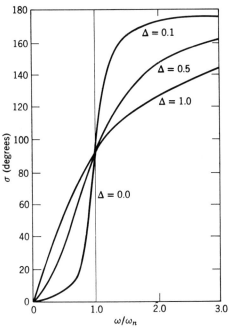

Figure 4.8. Phase angle for uncoupled heaving or pitching in waves.

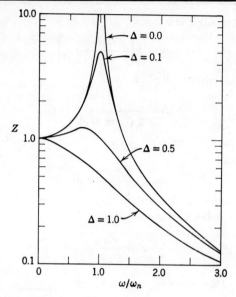

Figure 4.9. Magnification factor for uncoupled heaving or pitching in waves.

is called the *magnification factor* and is presented in Figure 4.9 as a function of the frequency ratio, ω/ω_n. The value of Z at *resonance* depends on the magnitude of the damping ratio, Δ, which in turn is a function of the damping wave amplitude ratio, $\bar{A}$, defined in Section 4.1C. Thus, as the value of $\bar{A}$ increases, the resonant response of the body decreases. One can conclude from this statement that projections on the submerged portion of a floating body that are close to the free surface, as sketched in Figure 4.10, will result in

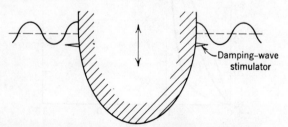

Figure 4.10. Hull projections to stimulate damping waves.

increased damping because of the larger-amplitude damping waves that these projections create. In addition to the wave damping, these projections increase the viscous-pressure damping. This is the reason for the use of *bilge keels* on surface ships to reduce the rolling motion.

The results presented in Figures 4.8 and 4.9 can also represent the *uncoupled pitching motions*. In this case the pitching response is

$$\theta = \theta_0 \cos(\omega t + \delta - \sigma)$$

$$= \frac{M_0/C}{\sqrt{(1 - \omega^2/\omega_n^2)^2 + [2\Delta\omega/\omega_n]^2}} \cos(\omega t + \delta - \sigma) \quad (4.53)$$

where the damping term is

$$\Delta = \frac{B}{2\sqrt{C(I_y + A)}} \quad (4.54)$$

the natural frequency is

$$\omega_n = \sqrt{C/(I_y + A)} \quad (4.55)$$

and

$$Z_\theta = \frac{\theta_0}{M_0/C} \quad (4.56)$$

is the magnification factor. It should be noted that the terminology used in the analysis of the uncoupled heaving and pitching motions is the same as that employed in vibration analysis [e.g., see Tong (1963)]. The analysis of the coupled motions is also similar to that of coupled vibrations; therefore, the reader is again referred to the book by Tong, or a similar text on vibrations, for a thorough treatment of equations like Eqs. 4.41 and 4.42.

In order to further clarify the nature of the motions of floating bodies, a symmetric body is studied experimentally in Section 4.3 and theoretically in Section 4.4.

4.2. MOORED AND TOWED BODIES

Many of the floating-body problems encountered by ocean engineers involve *mooring* or *towing cables*. As far as the mathematical descriptions of the motions of the floating bodies are concerned, the cable tension at the point of attachment to the body is simply an additional constraint in the equations of motion presented in Section 4.1. However, the engineer must be concerned also with the force and motions experienced by the cables themselves, since severe sea conditions can cause a cable to break, resulting in serious damage to the floating system.

Much of the significant work in the analysis of moored or towed systems has been carried out at the David Taylor Model Basin (now called the Naval Ship Research and Development Center) by Pode (1950, 1951), Wicker (1958), Springston (1967), and others. Kaplan and Raff (1969) and Kaplan (1970), in addition, present rather thorough treatments of mooring situations in both regular and irregular seas.

In this section an analysis of a *two-dimensional, steady, single-point moored system* is presented. Both extensible and inextensible cables are considered. Examples of moored systems are then analyzed and discussed.

The analysis begins by considering the cable tension at the point of attachment on an axially symmetric floating body in a uniform steady current, as shown in Figure 4.11. If static equilibrium is assumed to exist, the force relationships at the attachment point are the following: in the direction normal to the center line of the body

$$T_0 \sin(\phi_0 - \psi_0) = F_n + (W - F_b) \sin(\psi_0) \tag{4.57}$$

while in the tangential direction

$$T_0 \cos(\phi_0 - \psi_0) = F_t - (W - F_b) \cos(\psi_0) \tag{4.58}$$

The forces F_n and F_t are the hydrodynamic forces due to the viscous-pressure losses and the viscous (boundary-layer) losses, respectively, which are described in Section 1.5. These forces can be represented by

$$F_n = \tfrac{1}{2}\rho V_n^2 A_n C_n = \tfrac{1}{2}\rho V^2 \cos^2(\psi_0) A_n C_n \tag{4.59}$$

and

$$F_t = \tfrac{1}{2}\rho V_t^2 A_t C_t = \tfrac{1}{2}\rho V^2 \sin^2(\psi_0) A_t C_t \tag{4.60}$$

where A_n and A_t are the characteristic areas, and where C_n and C_t are the drag coefficients for the body.

4.2. MOORED AND TOWED BODIES

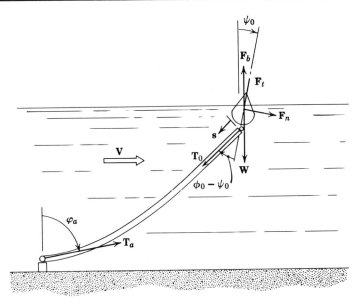

Figure 4.11. Moored symmetric buoy in a steady uniform current.

By dividing Eq. 4.57 by Eq. 4.58 the *angle between the tension*, T_0, *and the center line* of the body is found to be

$$\phi_0 - \psi_0 = \tan^{-1}\left[\frac{F_n + (W - F_b)\sin(\psi_0)}{F_t - (W - F_b)\cos(\psi_0)}\right] \quad (4.61)$$

Furthermore, the combination of Eqs. 4.57 and 4.58 obtained by eliminating the angle $(\phi_0 - \psi_0)$ yields the *tension at the point of attachment*, that is,

$$T_0 = \{F_n^2 + F_t^2 + 2(W - F_b)[F_n \sin(\psi_0) - F_t \cos(\psi_0)] + (W - F_b)^2\}^{1/2} \quad (4.62)$$

Thus ϕ_0 and T_0 are the boundary conditions for the cable. These boundary conditions are time dependent if either surface waves are encountered or vortex shedding occurs on the body and cable.

Now, consider an elemental segment of an inextensible mooring cable between the anchor and buoy sketched in Figure 4.12. Again,

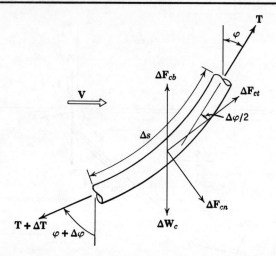

Figure 4.12. Element of an inextensible cable in a steady current.

if static equilibrium is assumed, the following force relationships result: in the axial direction of the cable at T_0

$$T = (T + \Delta T) \cos(\Delta\phi) - \Delta F_{ct} \cos\left(\frac{\Delta\phi}{2}\right) +$$

$$\Delta F_{cn} \sin\left(\frac{\Delta\phi}{2}\right) + \Delta W_c \cos(\phi) - \Delta F_{cb} \cos(\phi) \quad (4.63)$$

and in the normal direction

$$(T + \Delta T)\sin(\Delta\phi) = (\Delta W_c - \Delta F_{cb})\sin(\phi) + \Delta F_{ct}\sin\left(\frac{\Delta\phi}{2}\right)$$

$$+ \Delta F_{cn} \cos\left(\frac{\Delta\phi}{2}\right) \quad (4.64)$$

If the cable is of circular cross section, the hydrodynamic forces on the cable are those described in Section 1.5:

$$\Delta F_{ct} = \tfrac{1}{2}\rho V_t^2 \pi D C_t \, \Delta s = f_{ct} V_t^2 C_t \, \Delta s \quad (4.65)$$

and

$$\Delta F_{cn} = \tfrac{1}{2}\rho V_n^2 D C_n \, \Delta s = f_{cn} V_n^2 C_n \, \Delta s \quad (4.66)$$

The net weight (weight minus buoyancy) of the cable element can be expressed as

$$\Delta W_c - \Delta F_{cb} = (w_c - f_{cb}) \Delta s \qquad (4.67)$$

where w_c and f_{cb} are the cable *weight* and *buoyancy per unit length*.

Passing to the limit as $\Delta \phi \to 0$ and $\Delta s \to 0$ yields, for Eqs. 4.63 and 4.64, respectively,

$$dT = [f_{ct} V_t^2 C_t - (w_c - f_{cb}) \cos(\phi)] \, ds \qquad (4.68)$$

and

$$T \, d\phi = [(w_c - f_{cb}) \sin(\phi) + f_{cn} V_n^2 C_n] \, ds \qquad (4.69)$$

where the second-order terms are neglected, and where the results of Eqs. 4.65 through 4.67 have been included. The flow in both the axially symmetric boundary layer and the wake of the cable is assumed to be *turbulent*, so that (referring to Figure 1.8) the drag coefficients are approximately independent of the Reynolds number. Now, by eliminating ds, Eqs. 4.68 and 4.69 are combined to obtain the *differential equation for an inextensible cable*:

$$\frac{1}{T}\frac{dT}{d\phi} = \frac{f_{ct} V^2 \sin^2(\phi) C_t - (w_c - f_{cb}) \cos(\phi)}{(w_c - f_{cb}) \sin(\phi) + f_{cn} V^2 \cos^2(\phi) C_n}$$
$$= Q(\phi) \qquad (4.70)$$

Integration of Eq. 4.70 from the buoy, where $\phi = \phi_0$ and $T = T_0$, to any point on the cable results in the following:

$$\ln\left(\frac{T}{T_0}\right) = \int_{\phi_0}^{\phi} Q(\phi) \, d\phi \qquad (4.71)$$

Equation 4.71 can be integrated using the method of Pode (1951).

If the circular cross-sectioned cable is extensible, then (referring to Figure 4.13) *Hooke's law* yields the following expressions for the axial and radial strains, respectively:

$$\varepsilon_s = \frac{1}{E_c}\frac{T}{\pi R^2} = \frac{\Delta \delta}{\Delta s} \qquad (4.72)$$

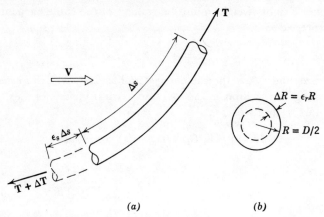

Figure 4.13. Element of an extensible cable in a steady current. (*a*) Axial elongation. (*b*) Radial contraction.

and

$$\varepsilon_r = -\frac{\mu}{E_c}\frac{T}{\pi R^2} = -\frac{\Delta R}{R} \ll 1 \qquad (4.73)$$

where E_c is *Young's modulus* for the cable material, R is the cable radius, and μ is *Poisson's ratio*. Note that in Eq. 4.73 the radial strain is rather small and, therefore, is neglected in this study. The hydrodynamic forces are, now,

$$\Delta F_{cn} = \tfrac{1}{2}\rho V_n^2 2(R - \Delta R)(1 + \varepsilon_s) C_n \,\Delta s$$
$$\simeq \tfrac{1}{2}\rho V_n^2 2R(1 + \varepsilon_s) C_n \,\Delta s \qquad (4.74)$$

and

$$\Delta F_{ct} \simeq \tfrac{1}{2}\rho V_t^2 \pi 2R(1 + \varepsilon_s) C_t \,\Delta s \qquad (4.75)$$

where terms of second order and higher are neglected. Since the total weight and buoyant force of the cable remain unchanged, the expression for the net weight of the elastic cable is the same as that for the inelastic cable.

When the results of Eqs. 4.74 and 4.75 are used, Eqs. 4.68 and 4.69, respectively, applied to an extensible cable are as follows:

$$dT = [f_{ct} V_t^2 C_t (1 + \varepsilon_s) - (w_c - f_{cb}) \cos (\phi)] \, ds \qquad (4.76)$$

and

$$T \, d\phi = [(w_c - f_{cb}) \sin (\phi) + f_{cn} V_n^2 C_n (1 + \varepsilon_s)] \, ds \qquad (4.77)$$

4.2. MOORED AND TOWED BODIES

The combination of Eqs. 4.76 and 4.77 by eliminating ds, while noting that $V_t = V \cos(\phi)$ and $V_n = V \sin(\phi)$, results in the *differential equation for the elastic cable*:

$$\frac{1}{T}\frac{dT}{d\phi} = \frac{f_{ct}V^2 \sin^2(\phi)C_t(1 + T/E_c\pi R^2) - (w_c - f_{cb})\cos(\phi)}{(w_c - f_{cb})\sin(\phi) + f_{cn}V^2 \cos^2(\phi)C_n(1 + T/E_c\pi R^2)}$$
$$= P(\phi, T) \tag{4.78}$$

where ε_s has been replaced by the expression in Eq. 4.72. The integration of Eq. 4.78 must be performed numerically since the variables ϕ and T cannot be separated.

Rather than pursuing the integration techniques required to solve the differential equations of the inelastic cable (Eq. 4.70) and the elastic cable (Eq. 4.78), attention is focused on two simplified (but very practical) cases involving the cable equations.

CASE 1

Consider the tension in either an elastic or an inelastic cable used to *taut-moor* a submerged body in water where no current is present. Since $V = 0$, there is no horizontal force on the cable or body, so that $\phi = 0$ over the entire cable length; thus, $\phi_0 = \psi_0 = 0$, as shown in Figure 4.14. In this case Eqs. 4.68 and 4.76 yield identical results

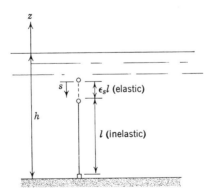

Figure 4.14. Comparison of taut-moored elastic and inelastic cables in a zero current.

since the hydrodynamic forces are equal to zero. The expression obtained for the tension at any point in the cable is, then,

$$T = T_0 + (f_{cb} - w_c)s \qquad (4.79)$$

where the tension at the attachment point on the body is obtained from Eq. 4.62, that is,

$$T_0 = W - F_b \qquad (4.80)$$

In this case the body must be *positively buoyant*; therefore the tension T_0 must be negative. From this result it is evident that the tension in the cable is equal to the sum of the net buoyant forces of the body and the cable above point s.

The *elongation* of a taut-moored elastic cable can be determined by combining the results of Eqs. 4.72, 4.79, and 4.80. Thus the differential elongation at point s is

$$\Delta\delta = \varepsilon_s \Delta s = \left(\frac{T}{E_c \pi R^2}\right) \Delta s = \frac{1}{E_c \pi R^2} [W - F_b + (f_{cb} - w_c)s] \Delta s$$

Passing to the limit as $\Delta s \to 0$ and integrating over the length of the cable yields the *total elongation*:

$$\delta = \int_0^\delta d\delta = \frac{1}{E_c \pi R^2} \int_0^l T\,ds = \frac{1}{E_c \pi R^2} \left[(W - F_b)l + (f_{cb} - w_c)\frac{l^2}{2}\right] \qquad (4.81)$$

CASE 2

In this case a *neutrally buoyant* inelastic cable, that is, one for which $f_{cb} = w_c$, is used to moor a relatively small float in a current of velocity V. Equation 4.70 applied to this situation becomes

$$\frac{1}{T}\frac{dT}{d\phi} = \frac{f_{ct}V^2 \sin^2(\phi)C_t}{f_{cn}V^2 \cos^2(\phi)C_n} = \pi\left(\frac{C_t}{C_n}\right)\tan^2(\phi)$$

Note that this result is the same as that obtained from the elastic equation (Eq. 4.78). The difference between the inelastic case and the elastic case appears in the boundary conditions at the float and the anchor, as shown in Figure 4.15. Integration of the previous equation yields the following result:

$$\ln\left(\frac{T}{T_0}\right) = \int_{\phi_0}^{\phi} \pi\left(\frac{C_t}{C_n}\right) \tan^2(\phi)\, d\phi$$

$$= \pi\left(\frac{C_t}{C_n}\right)[\tan(\phi) - \phi - \tan(\phi_0) + \phi_0] \quad (4.82)$$

Although the values of ϕ_0 and T_0 can be obtained from Eqs. 4.61 and 4.62, respectively, the determination of the cable angle and tension at any intermediate point requires an *a priori* knowledge of

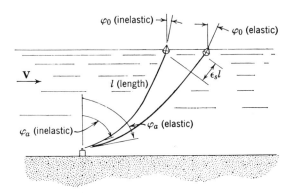

Figure 4.15. Comparison of neutrally buoyant elastic and inelastic cables in a steady uniform current.

the cable shape since without this knowledge one cannot determine the point, s, at which a particular value of ϕ occurs. A method used by Schick (1964) and discussed by Springston (1967) is to assume that the cable has a circular configuration, as illustrated in Figure 4.16. Assuming this configuration, one can see that at a known depth, d, and cable point, s, the following relations hold:

$$s = r(\phi - \phi_0) \quad (4.83)$$

and

$$d = r[\sin(\phi) - \sin(\phi_0)] \quad (4.84)$$

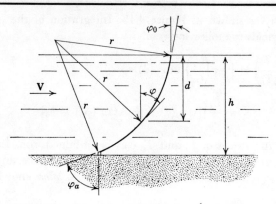

Figure 4.16. Circular-configured cable in a steady uniform current.

Equations 4.83 and 4.84 are combined by eliminating r to obtain

$$\phi - \phi_0 = \frac{s}{d}[\sin(\phi) - \sin(\phi_0)] \qquad (4.85)$$

Applying Eq. 4.85 to the *anchor point*, that is, where $\phi = \phi_a$ and $s = l$, yields

$$\phi_a - \phi_0 = \frac{l}{d}[\sin(\phi_a) - \sin(\phi_0)] \qquad (4.86)$$

where the length of the cable, the depth, and the value of ϕ_0 are known. Equation 4.86 must be solved numerically, using the technique described in Section 2.6c. By using this value of ϕ_a in Eq. 4.82 and letting $s = l$, the value of the curvature radius, r, can be determined:

$$r = \frac{l}{\phi_a - \phi_0} \qquad (4.87)$$

Using the value of r obtained in Eq. 4.87 in Eq. 4.83 yields the relation between s and ϕ. With this relationship the value of the tension at any point s on the cable can be obtained from Eq. 4.82.

The same method of analysis can be applied to determine the tension in a towing cable. Some towing situations that are encountered by ocean engineers are illustrated in Figure 4.17. These towing cable configurations and others involving inelastic cables are

4.3. EXPERIMENTAL STUDY OF BODY MOTIONS

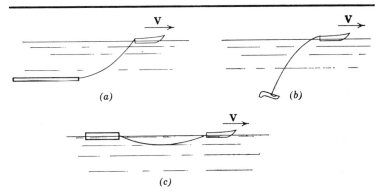

Figure 4.17. Some towing situations encountered by ocean engineers. (a) Towed geophysical line array. (b) Deep-towed active array. (c) Towed floating body using heavy cable.

analyzed by Pode (1951). The numerical techniques required in the analyses of situations involving elastic or extensible cables are described by Wang and Moran (1971).

One additional problem encountered in towing is due to the *twisting torque* of the cable. Cables are made of twisted strands of metal wire or synthetic fiber. When severe tensions are applied to these twisted or wound strands, they tend to straighten. This straightening can result in an unwanted rotation of the towed body due to the tension-induced torque in the cable. Moreover, repeated twisting and straightening of this type can result in *fatigue failure* of the cable.

Finally, the oscillatory motions of an elastic cable are analyzed by Kerney (1971), who uses the small-perturbation technique to derive the equations of motion. These equations are then integrated by using a numerical method.

4.3. EXPERIMENTAL STUDY OF BODY MOTIONS

In this section an experimental study of the *uncoupled heaving* and *pitching* motions of a body is described. The motions are uncoupled since the body is neither towed nor placed in a current. The purpose of the experiment is to illustrate both the experimental technique

148 FLOATING STRUCTURES IN WAVES

used in buoy motion studies and the method of experimental determination of several of the hydrodynamic coefficients described in Section 4.1.

The body used in the experiment is a pine circular cylinder having its axis of symmetry in the water plane, as sketched in Figure 4.18. In order to locate the center of rotation of the pitching motion at the center of gravity of the cylinder, a well is located about the midsection in which the heaving-pitching motions apparatus is attached. Also, two 0.8-lb weights are located in the well, one forward and one aft of the motions apparatus, in order to float the cylinder in a half-submerged position. With these weights and a 1.3-lb load due to the

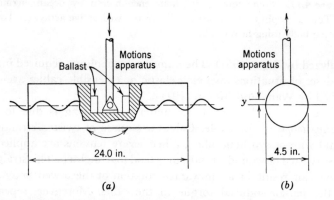

Figure 4.18. Schematic drawing of the heaving and pitching cylinder and motions apparatus. (*a*) Side view. (*b*) Front view.

motions apparatus, the *mass* of the heaving system is 0.214 slug, while the *moment of inertia* of the pitching body is 0.0193 lb-ft-sec^2.

Initially, several still-water measurements are made to determine both the natural frequencies and the system damping of the two degrees of freedom. To determine the values of these quantities the body is given an initial rectilinear or rotational displacement, depending on the motion in question, and then released, noting both the number of cycles over a given time and the amplitude decay over one period. The results obtained by following this procedure are shown in Figure 4.19, where the displacement of each degree of freedom is presented in terms of millivolts since the motions appa-

4.3. EXPERIMENTAL STUDY OF BODY MOTIONS

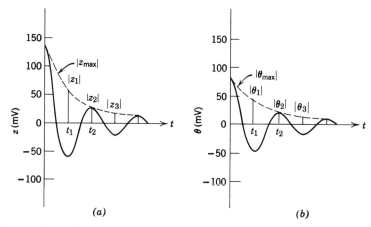

Figure 4.19. Damping and frequency results in calm water. (a) Heave. (b) Pitch.

ratus is not calibrated. The damped natural frequency values are found to be 1.70 Hz in heave and 1.74 Hz in pitch. For the damped free motions of the body the *decay* in the response amplitude is approximately exponential; for example, the absolute value of the *heaving amplitude* is given by

$$|z_{max}| = z_0 e^{-\Delta \omega_n t} \tag{4.88}$$

where Δ and ω_n are defined in Eqs. 4.50 and 4.51, respectively. Thus the ratio of two amplitudes at times t_1 and $t_3 = t_1 + T$ is

$$\frac{z_1}{z} = e^{\Delta \omega_n (t_1 - t_3)}$$
$$= e^{\Delta \omega_n (2\pi/\omega_d)} \tag{4.89}$$
$$= e^{2\pi \Delta / \sqrt{1-\Delta^2}}$$

since, according to Tong's (1963) book on vibrations, the *damped natural frequency* is defined by

$$f_d = \frac{\omega_d}{2\pi} = \frac{\omega_n}{2\pi} \sqrt{1-\Delta^2} \tag{4.90}$$

From the relationship represented in Eq. 4.89, the logarithmic decrement is found to be

$$\frac{2\pi\Delta}{\sqrt{1-\Delta^2}} = \ln\left(\frac{z_1}{z_3}\right) \qquad (4.91)$$

Using the values presented in Figure 4.19, one obtains a *damping ratio*, Δ, for the heaving motion of 0.184. Similarly, the Δ value for the pitching motion is 0.182. The values of the *undamped natural frequency*, f_n, are now obtained from Eq. 4.90, using the known values of Δ and f_d. Thus, for the heaving motions $f_n = 1.73$, while for the pitching motions $f_n = 1.75$. Comparing the damped and undamped natural frequency values, one sees that the damping has little effect on these frequencies.

Now, by knowing the values of Δ and ω_n and assuming the theoretical expression for the restoring coefficient presented in Eq. 4.38 to be accurate, so that

$$c = \rho g B' L = 46.8 \text{ lb/ft}$$

the hydrodynamic coefficients a and b described in Eq. 4.38 are obtained from the simultaneous solutions of Eqs. 4.50 and 4.51. Thus one finds $a = 0.191$ slug and $b = 1.64$ lb-sec/ft. Similarly, for the pitching motion the restoring coefficient obtained from Eq. 4.40 is

$$C = \frac{\rho g B' L^3}{12} = 15.6 \text{ lb-ft/rad}$$

The combination of Eqs. 4.54 and 4.55, therefore, yields $A = .0962$ lb-ft-sec^2 and $B = 0.49$ lb-ft-sec/rad. It is noted here that the theoretical expressions for the restoring coefficients c and C used in obtaining their respective values are based on the assumption that the body is wall-sided, that is, no body curvature exists at the waterline in the sectional plane. Since the body in question is cylindrical, this assumption is approximately valid only for small displacements. It is necessary to use the theoretical expressions since the restoring force-displacement (or moment-angular displacement) expression is nonlinear in the actual case. Thus the values of c and C actually vary with displacement.

4.3. EXPERIMENTAL STUDY OF BODY MOTIONS

After the calm-water measurements are made, the cylinder is subjected to deep-water waves of 2-in. height and various lengths. The amplitude responses of the body are shown in Figure 4.20 as a function of the wave (excitation) frequency. Again, since the motions apparatus is not calibrated, the amplitudes are presented in terms of millivolts. In Figure 4.20 one can observe that the peak response for the heaving motion occurs at 170 Hz, while that of the pitching motion occurs at 174 Hz, both frequencies being the damped natural frequencies of their respective degrees of freedom.

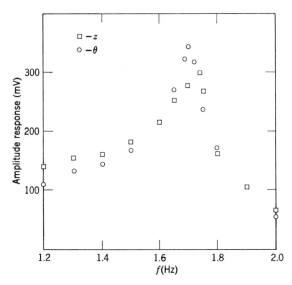

Figure 4.20. Amplitude responses of the heaving and pitching cylinder as a function of wave frequency.

During this experiment no attempt is made to separate the *viscous damping* and the *wave-making damping*. This can be done, however, by simply placing a wave probe, such as the resistance wire gauge described in Section 2.5, very near the body at a number of locations about the water plane during the calm-water measurements. In this way the amplitude of the wave, a_d, and that of the heaving body, A_0,

152 FLOATING STRUCTURES IN WAVES

can be monitored simultaneously. Thus the amplitude ratio $\bar{A} = a_d/A_0$ in Eq. 4.23 is known at a number of locations, and the wavemaking damping can be obtained by numerically integrating that equation. The difference between this value and that for the total damping is the viscous damping.

By using Eqs. 4.38 and 4.40, the theoretical values of the hydrodynamical coefficients a and A are found to be, respectively, 0.214 slug and 0.071 lb-ft-sec^2. Comparison of the experimental and theoretical values of the coefficient a shows excellent agreement, the experimental value being slightly lower. The reason for the lower experimental value is that the flow about the edges of the cylinder is three dimensional, resulting in a little less fluid mass being excited than for the two-dimensional flow. The agreement between the experimental and theoretical values of the added-mass moment of inertia coefficient, A, is not very good, the experimental value being 1.35 times that of the theoretical. The reason for this discrepancy is that the initial pitching displacement required to obtain an adequate electronic signal from the motion sensor is approximately 15° in the calm-water measurements. The damped free pitching motion is then nonlinear since Eq. 4.3 is no longer satisfied. The nonlinear pitching motion excites significantly more fluid mass than is excited by the linear motion.

4.4 ILLUSTRATIVE EXAMPLES

a. *Coupled Heaving and Pitching of a Fully Submerged Body*

A circular cylindrical oil tank 100 ft in length and 10 ft in diameter is being retrieved from the sea floor after being dumped from a barge in a severe storm. Some oil must first be pumped from the tank to reduce the ballast in order to make the tank neutrally buoyant. To avoid any severe weather or sea conditions the tank is then towed so that its axis of symmetry is 25 ft below the still-water level, as illustrated in Figure 4a. The towing speed is 1 knot, that is, 1.689 fps. The tank is towed into waves 3 ft in height and 100 ft in length. With this information and on the assumption that $I_y = 1.31 \times 10^7$

4.4. ILLUSTRATIVE EXAMPLES

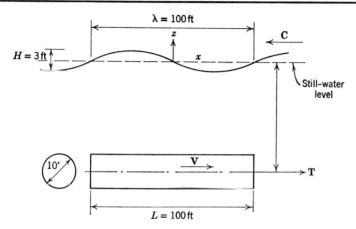

Figure 4a. Full submerged oil tank in waves.

lb-ft-sec², the heaving and pitching motions of the cylindrical tank are to be predicted, using the strip theory.

For the fully submerged circular cylinder the added-mass per unit length is found to be

$$m_w(\xi) = \rho \pi r^2 = 157 \text{ slugs/ft}$$

using the expression presented in Table 3.1. With this value, the added-mass and hydrodynamic moment of inertia coefficients in Eqs. 4.38 and 4.40 are, respectively,

$$a = 15{,}700 \text{ slugs}$$

and

$$A = 1.31 \times 10^7 \text{ lb-ft-sec}^2/\text{rad}$$

In this case the mass of the body is equal to the added-mass.

The tank experiences linear motions at a distance from the free surface, and, hence, the wave-making damping is negligible, so that the coefficient b is zero in the heaving equation (Eq. 4.37). Furthermore, since the body both is symmetric about the $y - z$ plane and has constant breadth, $B = 0$ in Eq. 4.39, the pitching equation. The coefficient e, however, is not equal to zero since it depends on the value of the added-mass; thus

$$e = -2Va = -5.31 \times 10^4 \text{ lb-sec/rad}$$

154 FLOATING STRUCTURES IN WAVES

The body is neutrally buoyant and fully submerged; therefore the hydrostatic restoring coefficients c and C in Eqs. 4.37 and 4.39, respectively, are both equal to zero. There is, however, a restoring moment due to the tension in the towing cable at the point of attachment. Referring to Figure 4b, one can write the expression for this restoring moment as

$$-\left(\frac{TL}{2}\right)\sin(\theta) \simeq -\frac{TL\theta}{2} = -1120\,\theta$$

where $R_L = 1.61 \times 10^7$, $C_t = 2.50 \times 10^{-3}$ and $T = 22.4$ lbs.
The value of 1120 lb-ft/rad, then, replaces the coefficient C in the pitching equation.

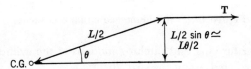

Figure 4b. Restoring moment due to cable tension.

The wave-induced force and moment expressions derived in Section 4.1D do not include the hydrostatic restoring terms in this situation since the body is fully submerged. Furthermore, the wave is a left-running wave, so that the displacement of the free surface is given by

$$\eta = a\cos(kx + \omega t)$$

Use of this expression in the forcing function expression of Eqs. 4.29 and 4.32 and the corresponding moment expressions simply involves changing the sign of the time term in Eq. 4.29 and that of $\sin(\omega t)$ in Eq. 4.32. The frequency in these expressions is not that of the wave, but that of the *frequency of encounter*; since the body is heading into the sea, the waves appear to approach the body at a speed of $V + c$ and, subsequently, a higher frequency. Therefore the frequency associated with this *apparent phase velocity* is

$$f_e = \frac{V+c}{\lambda} = \frac{\omega_e}{2\pi} = 0.243 \text{ Hz}$$

The integral

$$\int_{-B/2}^{B'/2} e^{kz}\,dy$$

appearing in the force and moment coefficients in Eqs. 4.38 and 4.40 is evaluated over the lower and upper halves of the body, independently for each half. In the case of Figure 4c, integration over the lower half of the body yields

$$\int_{-B'/2}^{B'/2} e^{kz_L}\, dy = \int_0^{\pi} 5 \sin(\alpha) e^{-25k - 5k \sin(\alpha)}\, d\alpha = 1.63 \text{ ft}$$

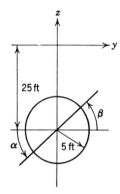

Figure 4c. Integration variables over the upper and lower halves.

following the method of solution of Eq. 4.34. Similarly, over the upper half of the cylindrical tank one obtains

$$\int_{-B'/2}^{B'/2} e^{kz_U}\, dy = -\int_0^{\pi} 5 \sin(\beta) e^{-25k + 5k \sin(\beta)}\, d\beta = -2.66 \text{ ft}$$

Using these values in the force coefficients of Eq. 4.38 and the moment coefficients of Eq. 4.40 yields the following results over both halves of the body:

$$F_A = F_B = M_A = 0$$

In addition to these results, over the lower half

$$M_{BL} = 1.79 \times 10^6 \text{ lb-ft}$$

while over the upper half

$$M_{BU} = -1.13 \times 10^6 \text{ lb-ft}$$

The equations of motion applied to the fully submerged towed oil tank are, then,

$$(m + a)\ddot{z} + e\ddot{\theta} = 0$$

and

$$(I_y + A)\ddot{\theta} + \frac{TL\theta}{2} = -(M_{BU} + M_{BL}) \sin(\omega_e t)$$

where the minus sign in the moment term is due to the left-running wave.

The solution of the pitching equation is

$$\theta = \frac{\dfrac{-(M_{BU} + M_{BL})}{TL/2} \sin(\omega_e t)}{1 - \dfrac{2(I_y + A)}{TL} \omega_e^2} = 1.08 \times 10^{-2} \sin(1.527t) \text{ rad}$$

where the natural (circular) frequency is

$$\omega_n = \sqrt{TL/2(I_y + A)} = 6.53 \times 10^{-3}/\text{sec}$$

The combination of the θ expression with the heaving equation yields

$$\ddot{z} = \frac{\dfrac{-(M_{BU} + M_{BL})\omega_e}{TL/2} \cos(\omega_e t)}{\left(\dfrac{m+a}{e}\right)\left(1 - \dfrac{2(I_y + A)}{TL} \omega_e^2\right)} = 2.79 \times 10^{-2} \cos(1.527t) \text{ ft/sec}^2$$

Upon integration the heaving displacement is found to be

$$z = \frac{-\ddot{z}}{\omega_e^2} = -1.20 \times 10^{-2} \cos(1.527t) \text{ ft}$$

Thus the cylindrical tank will experience little motion while towed at a depth of 25 ft in a 3-ft sea.

From the results presented in this example the following conclusions can be drawn:

1. A symmetric body with a constant cross section and a length that is an integer multiple of the wavelength will experience no net heaving force in either head or following seas. The body will

4.4. ILLUSTRATIVE EXAMPLES 157

experience only a pitching moment induced by the waves. This statement applies to both submerged and floating bodies.

2. The heaving and pitching motions of a submerged body decrease exponentially with depth when excited by waves.

3. Since the excitation frequency depends on the towing speed, resonance can be avoided by simply altering this speed. This does, however, involve changing the cable tension, which is also velocity dependent. For the circular cylinder considered in this problem the *axisymmetric boundary layer*, described in Section 1.5 and Example 1.6c, is almost entirely turbulent (referring to Figure 1.8). Thus the frictional resistance on the body is proportional to V^2, whereas the frequency of encounter, f_e, is proportional to V. This, means, then, that an increase in the towing velocity, V, results in a much greater change in the natural frequency of the body than in the frequency of encounter.

Plate 4.2. An artist concept of the Offshore Company's SCP-III Mark 2 featuring a semisubmerged hull form for stability in severe seas. (Courtesy of the Offshore Company, Houston, Texas.)

b. The Semisubmerged Hull: Effective Spring Constant of a Mooring Cable

The assembly sketched in Figure 4d is a section of a deep ocean floating platform. This section is towed to the station and then anchored in 1000 ft of water, using 890 ft of nylon cable. The neutrally buoyant cable has a diameter of 4 in. and a modulus of elasticity (Young's modulus), E_c, of 10^5 psi. The expressions for the heaving and pitching responses of the section are to be obtained by assuming that the horizontal element of the structure is always beneath the region of influence of the free surface.

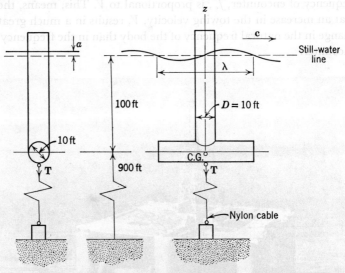

Figure 4d. A taut-moored semisubmerged structure.

Before analyzing the motions of the structure, the tension in the cable must be determined. The weight of the structure is 10^6 lb, while the weight of the displaced water is approximately 1.005×10^6 lb. The tension, T, in the cable is, then, 5000 lb. By using Eq. 4.72, the *longitudinal strain* in the cable is found to be

$$\varepsilon_s = \frac{\Delta l}{l} = \frac{1}{E_c} \frac{T}{\pi R^2} = 3.99 \times 10^{-3}$$

4.4. ILLUSTRATIVE EXAMPLES

Thus the *stretch* in the cable is

$$\Delta l = \varepsilon_s l = 3.55 \text{ ft}$$

This means that the actual depth of the axis of the horizontal cylinder is 96.45 ft. Since the difference in the actual depth and the design depth, that is, 100 ft, is less than 4 ft, the design conditions shown in Figure 4d are used in the motions analysis. Furthermore, since the cable is both elastic and under tension, an *effective spring constant* is needed in the equations of motion. This constant can be obtained by rearranging the longitudinal strain expression to obtain

$$k_c = \frac{T}{\Delta l} = \frac{E_c \pi R^2}{l} = 1410 \text{ lb/foot}$$

The added-mass per unit length, $m_w(\xi)$, is obtained for the horizontal cylinder only, since the linear motion of the vertical cylinder is primarily in its axial direction. Furthermore, the added-mass coefficient, a, in Eq. 4.38 must be obtained by breaking the integral into parts, since $m_w(\xi)$ on the portion of the horizontal cylinder beneath the vertical cylinder is approximately one-half the value of $m_w(\xi)$ for the end sections. with reference to Figure 4e,

$$a = \int_5^{50} \rho \pi r^2 \, d\xi + \int_{-5}^{5} \frac{\rho \pi r^2}{2} \, d\xi + \int_{-50}^{-5} \rho \pi r^2 \, d\xi =$$

$$1.49 \times 10^4 \text{ lb-sec}^2/\text{ft}$$

where the added-mass expression for the circular cross section is obtained from Table 3.1.

Similarly, the *hydrodynamic moment of inertia coefficient*, A, must also include the effect of the rotating vertical cylinder in addition to the effect of the horizontal cylinder's rotation. The expression for A in Eq. 4.40 is, then,

$$A = \rho \pi r^2 \left(\int_5^{50} \xi^2 \, d\xi + \int_{-5}^{5} \frac{\xi^2}{2} \, d\xi + \int_{-50}^{-5} \xi^2 \, d\xi + \int_5^{100} \chi^2 \, d\chi \right)$$

$$= 6.55 \times 10^7 \text{ lb-ft-sec/rad}$$

where χ is the vertical coordinate from the center of gravity, as shown in Figure 4e.

160 FLOATING STRUCTURES IN WAVES

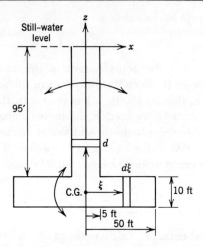

Figure 4e. Pitching geometry.

Model tests of the section indicate that the *damping ratio*, $\bar{A}$, in Eqs. 4.23 and 4.24 is approximately $\frac{1}{5}$ in value. The linear *damping coefficient*, b, corresponding to this value is

$$b = \frac{\pi \rho g^2}{8\omega_n^3} \int_{-5}^{5} \bar{A}^2 \, d\xi = \frac{326}{\omega_n^3}$$

where, since the body is excited by surface waves, ω_n is equal to the wave frequency. The corresponding damping coefficient, B, in the pitching equation is

$$B = \frac{2720}{\omega_n^3}$$

since the current velocity is zero.

The *linear restoring coefficient*, c, in this case depends on both the *buoyancy* of the body and the *spring constant* of the neutrally buoyant mooring cable. The restoring force due to the cable can be expressed as

$$k_c z = 1410 z$$

Thus, from Eq. 4.38,

$$c + k_c = \rho g \int_{-5}^{5} B'(\xi) \, d\xi + k_c = \rho g \int_{-5}^{5} 2\sqrt{25 - \xi^2} \, d\xi + k_c$$
$$= 6470 \text{ lb/ft}$$

4.4. ILLUSTRATIVE EXAMPLES

The *restoring moment* is also affected by the cable tension, the additional moment being

$$T[5 \sin (\theta)] \simeq T (5\theta) \simeq k_c\left(z + \frac{5\theta^2}{2}\right)5\theta$$

referring to Figure 4*f*. This component of the restoring moment is negligible for small (linear) heaving and pitching displacements. The *restoring moment coefficient*, C, in Eq. 4.40 is, then,

$$C = \rho g \int_{-5}^{5} \xi^2 B'(\xi) \, d\xi = 31{,}400 \text{ lb-ft/rad}$$

With reference to Case 6 in Section 4.1E, the body is symmetric with a constant damping amplitude ratio, that is, $\bar{A} = \frac{1}{5}$, and is experiencing no current. Therefore, $d = e = h = D = E = H = 0$.

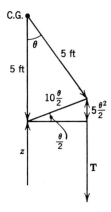

Figure 4*f*. Restoring moment due to cable tension.

The wave-induced force and moment result from the changes in buoyancy of the vertical cylinder only. It should be noted, however, that the force due to the horizontal velocity component of the particles within the wave can be significant for long waves. This force is taken into account if the surging motion is analyzed. This horizontal force also results in an additional pitching moment. The inclusion of this force and the resulting pitching moment into the analysis of the motions of floating bodies is treated in the report by

Kaplan and Raff (1969). The wave-induced force and moment coefficients described by Eqs. 4.38 and 4.40, respectively, which result from the changes in buoyancy are the following:

$$F_A = \rho g a_w \int_{-5}^{5} B_V' \cos(k\xi)\, d\xi = \rho g a \int_{-5}^{5} 2\sqrt{25 - \xi^2} \cos(k\xi)\, d\xi$$

$$= 2\rho g a_w \int_{-5}^{5} \sqrt{25 - \xi^2} \left[1 - \frac{(k\xi)^2}{2!} + \frac{(k\xi)^4}{4!} - \cdots \right] d\xi$$

$$= 5.05 \times 10^3 a(1 - 3.12k^2 + 3.25k^4 - \cdots)$$

$$F_B = 0$$

$$M_A = 0$$

$$M_B = 3.16 \times 10^4 a_w (1 - 2.08k^2 + 1.63k^4 - \cdots)$$

The equations of motion are, now,

$$(m + a)\ddot{z} + b\dot{z} + (c + k_c)z = F_A \cos(\omega t)$$

and

$$(I_y + A)\ddot{\theta} + B\dot{\theta} + C\theta = M_B \sin(\omega t)$$

in heave and pitch, respectively. These equations are in the same form as the uncoupled equations, Eqs. 4.46 and 4.47.

The *undamped natural frequency* in heave is obtained from Eq. 4.51 and found to be

$$f_{nz} = \frac{\omega_{nz}}{2\pi} = \frac{1}{2\pi}\sqrt{(c + k_c)/(m + a)} = 0.0608 \text{ Hz}$$

while that for the pitching motion is

$$f_{n\theta} = \frac{\omega_{n\theta}}{2\pi} = \frac{1}{2\pi}\sqrt{C/(I_y + A)} = 0.00242 \text{ Hz}$$

where $I_y = 7 \times 10^7$ lb-ft-sec.

A deep-water wave having a frequency equal to the natural frequency in heave has a length of

$$\lambda = \frac{gT^2}{2\pi} = 1390 \text{ ft}$$

from Eq. 2.25. This wave is significantly larger than the design wave, which is 190 ft (twice the depth of the uppermost part of the horizontal cylinder). Similarly, the resonant wave for the pitching motion would be an order of magnitude larger than the resonant wave of the heaving motion. From these results one sees that the structure will experience little motion for all but extremely long waves.

The type of structure discussed in this example is commonly referred to as the *semisubmerged structure*. It is used in both working platforms and vehicles, such as that described by Lang (1971), since the structure is relatively insensitive to surface waves, as illustrated in the example.

c. Computer Analysis of Uncoupled Buoy Motions

The phase angle, σ, and the magnification factors, Z and Z_θ, are easily calculated using the digital computer. To establish curves of these parameters, such as those presented in Figures 4.8 and 4.9 for the heaving motion, a program for the solution of Eqs. 4.49, 4.52, and 4.56 must be created by following a flow chart like that shown in Figure 4g for the calculation of σ and Z as functions of the frequency ratio ω/ω_n. The resulting FORTRAN program is as follows:

```
         READ 100, A, B, C, G, M, PI
         DEL = B/(2*SQRT(C*(M + A)))
         ØMAN = SQRT(C/(M + A))
         N = 1
    10   RAØM = N/100
         ØMA = RAØM*ØMAN
         LAM = 2.*PI*G/(ØMA**2)
         PRINT 200, RAØM, ØMA, LAM
         Z = 1./SQRT((1. − RAØM**2)**2 + (2.*DEL*RØAM)**2)
         SIGMA = ATAN(2.*DEL*RØAM/(1. − RØAM**2))
         PRINT 300, Z, SIGMA
         N = N + 1
         IF(RØAM − 3)20, 20, 30
    20   GØ TØ 10
    30   STØP
    100  FØRMAT (6E12.6)
    200  FØRMAT (3E12.7)
    300  FØRMAT (2E12.6)
```

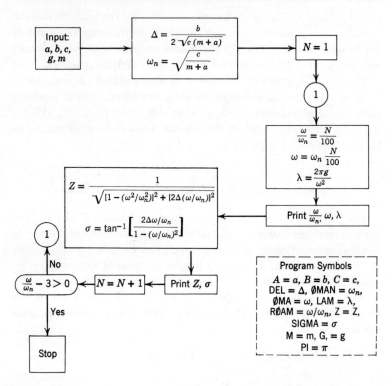

Figure 4g. Flow chart for the computation of the magnification factor and phase angle for uncoupled heaving motion.

Programs have been developed at institutions such as the Naval Ship Research and Development Center in Carderock, Maryland, that can be used to compute the values of the hydrodynamic coefficients and the vehicle responses for various hull forms. Furthermore, the equations of motion can also be programmed for an analog computer to obtain continuous-time responses once the coefficients are evaluated. The most recent advancement in computer-aided motions analyses is the development of the hybrid computer, which combines the speed and accuracy of the digital computer with the continuous aspect of the analog computer.

4.5. REFERENCES

Blagoveshensky, S. (1962), *Theory of Ship Motions*, Dover Publications, New York.

Comstock, J., ed. (1967), *Principles of Naval Architecture*, Society of Naval Architects and Marine Engineers.

Grim, O. (1959), "Die Schwingungen von Schwimmenden, Zweidimensionalen Körpern," Hamburgische Schiffbau-Versuchsanstalt *Gesellschaft Rept.* No. 1771, September.

Ince, E. (1956), *Ordinary Differential Equations*, Dover Publications, New York.

Jacobs, W., et al. (1960), "Guide to Computational Procedure for Analytical Evaluation of Ship Bending-Moments in Regular Waves," *Stevens Inst. Tech. (Davidson Lab.) Rept.* No. 79, October.

Kaplan, P. (1970), "Hydrodynamic Analysis Applied to Mooring and Positioning of Vehicles and Systems in a Seaway," 8th Symposium on Naval Hydrodynamics, Pasadena, Calif.

Kaplan, P., and Raff, I. (1969), "Development of a Mathematical Model for a Moored Buoy System," *Oceanics, Inc., Tech. Rept.* No. 69-61, April.

Kerney, K. (1971), "Small-Perturbation Analysis of Oscillatory Tow-Cable Motion," *Naval Ship Res. Develop. Center Rept.* No. 3430, November.

Lamb, H. (1945), *Hydrodynamics*, Dover Publications, New York.

Lang, T. (1971), "S^3—New Type of High-Performance Semisubmerged Ship," *Am. Soc. Mech. Engrs., Paper* No. 71-WA/UnT-1.

Lewis, F. (1929), "The Inertia of the Water Surrounding a Vibrating Ship," *Trans. Soc. Naval Architects Marine Engrs.*

Pode, L. (1950), "A Method of Determining Optimum Lengths of Towing Cables," *David Taylor Model Basin, Rept.* No. 717, April.

Pode, L. (1951), "Tables for Computing the Equilibrium Configuration of a Flexible Cable in a Uniform Stream," *David Taylor Model Basin, Rept.* No. 687, March.

Schick, G. (1964), "Design of a Deep-Moored Oceanographic Station," Marine Technology Society, Buoy Technical Symposium.

Springston, G. (1967), "Generalized Hydrodynamic Loading Functions for Bare and Faired Cables in Two-Dimensional Steady-State Cable Configurations," *David Taylor Model Basin Rept.* No. 2424, June.

Tong, K. (1963), *Theory of Mechanical Vibrations*, John Wiley, New York.

Wang, T., and Moran, T. (1971), "Analysis of the Two-Dimensional Steady-State Behavior of Extensible Free Floating Cable Systems," *Naval Ship Res. Develop. Center Rept.* No. 3721, October.

Wendel, K. (1956), "Hydrodynamic Masses and Hydrodynamic Moments of Inertia," *David Taylor Model Basin Transl.* No. 260, July.

Wicker, L. (1958), "Theoretical Analysis of the Effect of Ship Motion on Mooring Cables in Deep Water," *David Taylor Model Basin Rept.* No. 1221, March.

Wylie, C. (1960), *Advanced Engineering Mathematics*, McGraw-Hill, New York.

4.6. PROBLEMS

1. A box-shaped barge 100 ft in length, 20 ft wide, and 10 ft high draws 5 ft of salt water ($\rho = 2.00$ slugs/ft^3). Determine the hydrodynamic coefficients, the critical damping value, and the values of the natural and damped natural frequencies. Assume that $\bar{A} = 0.1$, $V = 0$, and the $m_w(\xi)$ value is half that of the $A/B = 2$ rectangle of Table 3.1.

2. If 2-ft waves, which are 50 ft in length, are present in the situation described in Problem 1, determine the force coefficients, F_A and F_B, the moment coefficients, M_A and M_B, and the expressions for the magnification factors of each degree of freedom.

3. A neutrally buoyant and fully submerged body having the dimensions of the barge described in Problem 1 has its center of gravity at the geometric center, which is 25 ft below the free surface. The length and beam of the body are parallel to the still-water level. Determine the heaving and pitching equations if the body experiences 2-ft deep-water waves that are 50 ft long. Assume a current velocity of 1 fps. *Note:* z in the force and moment expressions is at the top and bottom of the body.

4. Determine the expressions for the magnification factors for both heaving and pitching in Problem 3.

5. A circular cylindrical spar buoy having a hemispherical bottom, as shown in the figure, experiences waves 100 ft in length and

2 ft in height. Determine the equations of motion in heave and pitch assuming that $\bar{A} = 0.10$.

6. Determine the values of the natural frequencies and the magnification factors at resonance for each degree of freedom in Problem 5.

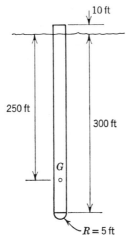

Problem 5. A spar buoy.

7. The semisubmerged hull is a rather stable structure in moderately heavy seas since most of the hull is below the region of wave action and the primary disturbing force is, therefore, the buoyancy change on the legs or struts.

For the structure in Example 4.4b determine the expressions for both the restoring force and moment if the single mooring cable attached to the center of the submerged hull is replaced by two similar cables attached at each end of the hull, that is, at $\xi = +50$ and $\xi = -50$.

8. Determine the natural frequencies of the heaving and pitching motions of the body in Problem 7.

9. A geophysical line array, 100 ft in length and 6 in. in diameter is towed by a ship traveling at a speed of 10 knots (1 knot = 1.689 fps). The array is neutrally buoyant and travels in a horizontal plane. At the point of attachment to the array the

towing cable is horizontal, while at ship the cable is at a 45° angle to the horizontal. Assuming a turbulent boundary layer to exist over the entire array and also about the cable, determine both the cable tension at the ship and the net weight distribution of the cable. The cable is 1 in. in diameter and 1,000 ft in length. Refer to Section 1.5 for the method of computation of the drag on the array. Assume that the drag coefficients C_n and C_t for the cable have the values at the cable midpoint and, in addition, assume that the cable profile is circular, Hint: Use equations (4.68) and (4.69), where $\Delta s = R \Delta \phi$.

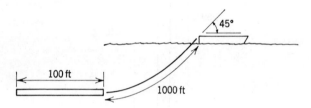

Problem 9. A towed geophysical line array.

10. The seismic array in Example 1.6c is towed, using a neutrally buoyant 1-in. diameter, elastic cable. The cable assumes a circular configuration and is approximately horizontal at the towing ship. Assuming the values of C_n and C_t to be those at the cable midpoint (obtained from Figure 1.8), determine the tension in the cable at the ship and also the strain in the cable. The cable length is 500 ft. Hint: Use equations (4.76) and (4.77), where $\Delta s = R \Delta \phi$.

Appendix A

FORTRAN IV LANGUAGE

1. Algebraic Symbols:

    +   Addition, e.g., $a + b \rightarrow A + B$
    −   Subtraction, e.g., $a - b \rightarrow A - B$
    *   Multiplication, e.g., $ab \rightarrow A*B$
    /   Division, e.g., $a/b \rightarrow A/B$
    **   Exponent, e.g., $a^2 \rightarrow A**2$.

2. Variables:

    I, J, K, L, M, N   These are fixed point variables. Floating point variables may not begin with any of these letters. All other letters can be used to begin floating point variables. All variables must begin with letters.

3. Functions:

    ABS( )   Absolute value of the bracketed term, e.g., $|a| \rightarrow$ ABS(A)

    LØG( )   Natural logarithm of the bracketed term, e.g., $\ln(v) \rightarrow$ LØG(V)

    SQRT( )   Square root of the bracketed term, e.g., $\sqrt{b} \rightarrow$ SQRT(B)

EXP( )   Exponential function of the bracketed term, e.g., $e^c \to \text{EXP(C)}$

SIN( )   Sine of the bracketed angle, e.g., $\sin(\theta) \to \text{SIN(THETA)}$

CØS( )   Cosine of the bracketed angle, e.g., $\cos(\theta) \to \text{CØS(THETA)}$

ATAN( )   Arc tangent of the bracketed term, e.g., $\tan^{-1}(e) \to \text{ATAN(E)}$

4. FØRMAT Statement:

   n FØRMAT (2E15.6)   "n" is the statement number. This statement says that two variables (2) are floating point numbers with exponents (E) using a total of fifteen (15) spaces with eight (6) spaces on the right-hand side of the decimal point. The letter F is used for floating point numbers without exponents, and I is used to indicate integers.

5. Input and Output Statements

   READ n, A, B   Read into the computer the values of A and B following the FORTRAN statement in statement number n.

   PRINT n, A, B   Print out the values of A and B following the FORTRAN statement in statement number n.

6. Control Statements

   GØ TØ n   Execute statement n next and the sequence following statement n.

   IF( ) n_1, n_2, n_3 If the bracketed term is negative, go to statement n_1; if the term is zero, go to statement n_2; if the term is positive, go to statement n_3.

   STØP   This term stops the computer at any point in the program.

   END   This statement is used when the program is completed.

Appendix B

SHOALING WAVE TABLES

h/λ	$c/c_0, \lambda/\lambda_0$	H/H_0
0.004	0.02503	4.47016
0.00564	0.03548	3.7546
0.00691	0.0434	3.3951
0.00798	0.05012	3.15972
0.00892	0.05604	2.98854
0.00978	0.06132	2.85725
0.01057	0.06624	2.74941
0.01129	0.07088	2.65825
0.01198	0.07516	2.58175
0.01263	0.07917	2.51575
0.01788	0.11183	2.11892
0.02193	0.13682	1.91764
0.02534	0.15784	1.78731
0.02836	0.17633	1.69276
0.03109	0.19298	1.61979
0.03363	0.20816	1.56124
0.03598	0.22234	1.51224
0.03821	0.23551	1.47087

(*Continued*)

SHOALING WAVE TABLES

(continued)

h/λ	$c/c_0, \lambda/\lambda_0$	H/H_0
0.04031	0.24807	1.43469
0.05764	0.34701	1.22586
0.07135	0.42047	1.12545
0.08329	0.48022	1.0643
0.09415	0.53107	1.02284
0.10431	0.57522	0.99322
0.11394	0.61434	0.97128
0.1232	0.64934	0.95473
0.1322	0.68081	0.94221
0.14098	0.70933	0.93273
0.22512	0.88841	0.91807
0.31212	0.96117	0.94894
0.40496	0.98774	0.97602
0.50183	0.99636	0.99049
0.60063	0.99895	0.99657
0.70021	0.9997	0.99883
0.80007	0.99991	0.99961
0.90002	0.99998	0.99987
1.00001	0.99999	0.99996
2	1	1
3	1	1
4	1	1
5	1	1

Appendix

C

ANSWERS TO SELECTED PROBLEMS

Chapter 1

1. $t = 1.23$ years
2. $b = 11.1$ ft
3. $d = 124$ ft

5. (a) $\psi = RV_0 \left(\dfrac{r}{R} - \dfrac{R}{r} \right) \sin(\theta)$

   $= 10 \left(r - \dfrac{1}{r} \right) \sin(\theta)$

   (c) At $r = R$ and $\theta = \pi/2$, $V = 2V_0 = 20$ fps
   (d) $V_0 = 25.8$ fps

6. $F_t = 78.6$ lb

Chapter 2

1. (a) $\lambda = 35.5$ ft and $c = 8.03$ fps
   (b) $E = 20{,}100$ lb-ft/ft, $\dot{E} = 2{,}270\,\mathbf{i}$

173

174 APPENDIX

2. $\psi = -\dfrac{ag}{\omega}\dfrac{\sinh(kh+kz)}{\cosh(kh)}\cos(kx)\sin(\omega t)$

3. $H = 0.951$ ft in cold water
4. $H = 3.18$ ft
5. (a) $\lambda = 1.57$ ft;
   (b) $h = 0.25$ ft

6. $a = \dfrac{1}{k}\dfrac{\sinh(kh)}{\cosh(kh+ka)}$

7. (a) $t = 4.65$ min
   (b) $L = 1{,}000$ ft
9. $\lambda = 253$ ft
11. $\Omega = -0.0718/\text{sec}$

12. $\eta = -\dfrac{A\omega J_0(kr)}{g}\cosh(kh+kz)\cos(\omega t)$

13. As ρ_0 increases, k increases and the wave energy, therefore, is reduced; thus a "calming" does occur.

Chapter 3

2. $F(t) = 308e^{0.943\cos(4.50t)} - 72.4\cos^2(4.50t) + 1160$
   $M(t)\}^+ = -489\cos(4.50t)\{2.77e^{0.943\cos(4.50t)}$
   $\qquad + 0.943\cos(4.50t)e^{0.943\cos(4.50t)}\} + 72.8\cos^3(4.50t)$
   $\qquad + 435\cos^2(4.50t) - 36.0$

3. (a) $w_{\max} = 0.760$ fps
   (b) $u_{\max} = 0.829$ fps
5. $\lambda = 62.9$ ft, $\lambda_0 = 82.6$ ft, $c = 15.7$ fps, and $c_0 = 20.7$ fps
6. 26.4 ft offshore
7. $\beta = 32.5°$

8. $v_n = \dfrac{ag}{c}e^{kz}\{\cos(kx-\omega t)\sin(\alpha) - \sin(kx-\omega t)\cos(\alpha)\}$

   $v_t = \dfrac{ag}{c}e^{kz}\{\cos(kx-\omega t)\cos(\alpha) + \sin(kx-\omega t)\sin(\alpha)\}$

9. $F_{\max} = 34.5$ lb, $M_{\max} = 977$ lb-ft

10. (a) $w(z, t) = -\frac{1}{2}\rho DC_n(ag/c)^2 \cos(\omega t)|\cos(\omega t)|$
$+ \frac{1}{4}\rho\pi D^2 C_i \, agk \sin(\omega t) = w(t)$
(b) $F_s(z, t) = w(t)(\eta - z)$, since $F_s(\eta, t) = 0$
(c) $M_b(z, t) = -\frac{w(t)}{2}(\eta - z)^2$, since $M_b(\eta, t) = 0$
(d) $X(z, t) = \frac{w(t)}{6EI}\left\{-\frac{(\eta - z)^4}{4} - (\eta + h)^3(h + z) + \frac{(\eta + h)^4}{4}\right\}$, since $X = \frac{\partial X}{\partial z} = 0$ at $z = -h$

11. Wave frequency $= 0.506$ Hz, vortex shedding frequency $= 1.47$ Hz, $f_1 = 2.55$ Hz; thus near resonance with vortex shedding occurs
12. (a) $f_1 = 4.89$ Hz
    (b) braces applied at $0.63\,L$ from the base
13. (a) $f_v = 1.53$ Hz
    (b) $f_v = 4.89$ Hz
14. $\lambda_t = 55.7$ fy
15. (a) $\alpha = 19°28'$
    (b) $F_R = 406$ lb
    (c) $F_{Rt} = 312$ lb
    (d) $F_{Rd} = 93.5$ lb

Chapter 4

1. $a = 4.27 \times 10^4$ lb-sec²/ft
$b = 846$ lb-sec/ft
$c = 1.28 \times 10^5$ lb/ft
$d = e = h = 0$
$A = 3.56 \times 10^7$ lb-ft-sec²
$B = 7.30 \times 10^5$ lb-ft-sec
$C = 1.07 \times 10^8$ lb-ft
$D = E = H = 0$
Critical damping $= 1.79 \times 10^5$ lb-sec/ft in heave
$f_n = 0.2274$/sec in heave and pitch
$f_d = 0.2269$/sec in heave and pitch

2. $F_A = F_B = M_A = 0$, and $M_B = -1.57 \times 10^5$ lb-ft
   $Z_\theta = \{(1 - 0.490\omega^2)^2 + 6.81 \times 10^{-3}\omega^2\}^{-1/2}$
3. Heaving: $12.6 \times 10^4 \ddot{z} - 171\ddot{\theta} = 0$
   Pitching: $10.5 \times 10^7 \ddot{\theta} = -2.12 \times 10^5 \sin(0.125t)$
4. $Z = Z_\theta = 0$, since the natural frequencies for each degree of freedom are zero
7. $F_2 = -7850z$, and $M_2\}^+ = -7.02 \times 10^6 \, \theta$
8. Heaving frequency $= 0.0668$ Hz
   Pitching frequency $= 0.00115$ Hz
9. $T_0 = 2321$ lb and $(w_c - f_{cb}) = 1.83$ lb/ft

INDEX

Added-mass, 80–81, 82, 85–86, 116, 118, 119, 150, 152, 153, 159
Added-mass coefficient, 80–81, 82, 85–86
Anchor point, 146

Beating, 97
Bending moment, 111
Bernoulli's equation, 11, 65, 68
Bessel function, 55
Bessel's equation, 55
Bilge keel, 137
Boundary layer, 12
Bow wave, 89
Braced structure, 87
Breaking wave, 34, 47, 74–75
Buoy, spar, 131

Cable elongation, 144
Cauchy-Riesmann equations, 8, 22
Cavitation, 20, 22
Celerity, 25, 31, 34, 43, 46, 52, 58
Circular cable configuration, 145–146
Circular wave frequency, 29, 58
Circulation, 5, 16, 19
Compressibility of seawater, 5
Continuity, equation of, 3–4, 28
Contour line, 76
Convection velocity, 47
Cramer's rule, 134
Critical damping, 135
Cylinder, circular, floating, 147–152
  flow past, 12–14, 78–83
  submerged, 152–157

Damped natural frequency, 149
Damping ratio, 135, 137, 150
Damping, wave, 121–124, 151–152
  viscous, 151–152
Deep water wave, 33
Differential operators, 133
Divergent waves, 89
Drag, wave making, 89–94, 98–100
Drag coefficient, 13, 19, 79–80, 99, 138, 140–141
Dynamic free-surface condition, 26, 27

Encounter, frequency of, 154
Energy, wave, 37–40
  kinetic, 38–39
  potential, 38
  spectrum, 85–86
Energy flux, 39–40, 91, 122
Extensible cable, 141–145, 158–163

Fatigue failure, mooring line, 147
Frequency, wave, 29, 58
Frequency of occurrence, 85–86
Froude-Krylov hypothesis, 113, 125
Froude number, 91–92, 94–95, 99, 107

Geophysical line array, 147
Group, wave, 36–37, 91
Group velocity, 36

Heaving motion, 112–113, 130–131, 133–137, 147–164
Hooke's law, 141

177

INDEX

Hydrodynamic coefficients, 130–131, 132–133
Hydrodynamic moment of inertia, 159
Hydrofoil lift, 16, 19
Hydrostatic equation, 3
Hydrostatic theorem, 3

Inextensible cable, 139–141
Intermediate depth wave, 34
Irrotational flow, 7–8, 11, 25, 65, 125

Kelvin wave pattern, 90–94
Kinematic free-surface condition, 26
Kinetic energy, 38–39
Knot, 21

Laplace's equation, 28, 54
    cylindrical coordinates, 54
Left-running wave, 30
Lift force, 16, 19
Linearized free-surface condition, 28, 29
Littoral drift, 77–78
Littoral transport, 77–78
Loading function, 111
Logarithmic decrement, 150
Longitudinal strain, 158

Magnification factor, 135–136, 137
Moored bodies, 137–147, 158–163
Motion, equations of, 114, 130–132, 133, 134

Natural frequency, damped, 149
    fixed structures, 85–89, 95–98
    floating structures, 122–124, 135–137
Navier-Stokes equations, 10
Neumann, function, 56
Neutrally buoyant body, 152–157
Neutrally buoyant cable, 144–146

Orthogonal, 76

Pitching motion, 112–113, 114, 130–131, 133–134, 137, 147–164
Poisson's ratio, 142
Potential energy, 38
Power, 122
Power transmission, 73
Pressure residue, 65–67

Reflection, wave, 67
Refraction, wave, 75–76, 103–107
Refraction coefficient, 104
Resistance wire wave gauge, 50–52, 96
Resonance, 89, 103, 136
Reynolds number, 13, 19, 79–80, 103, 14
Right-running wave, 30
Rolling motion, 112–113

Sand ripples, 77
Seafloor condition, 26, 28–29
Seawater, 19
Semisubmerged structure, 158–163, 167
Separated flow, 12
Shallow water wave, 34–35
Shear force, 111
Shoaling waves, 73–76, 103–107
Sphere, floating, 119–130, 121, 127–129
Spilling wave, 35
Spring constant, effective, 159
Standing wave, 29, 67–71
Stokes' law of viscosity, 10
Stokes' theory, 41–47
    first order, 44–45
    second order, 46
Streamfunction, 7–8, 22
Strip theory, 113, 115
Strouhal number, 88, 103
Successive approximations, method of, 59
Surging motion, 112–113
Swaying motion, 112–113

Taut-morred body, 143, 158–163
Transverse waves, 89
Traveling wave, 30
Trim angle, 115
Trochoidal theory, 48–50
Trochoidal wave, 48–50
Twisting torque, 147

Unbraced structure, 83–85, 87, 95–98

Velocity potential, 6–7, 29, 30, 43, 46, 6
Viscosity, coefficient of, 10, 19
Vortex shedding, 87–89, 100, 103, 139
Vorticity, 49–50

Wake, 12
Wall-sided body, 120, 121, 150

Wave-making drag, 89–94, 98–100
Wave number, 28, 31, 34
Wave properties, 25, 31, 34, 43, 46, 59

Yawing motion, 112–113
Young's modulus, 141–143, 158